SHE WHO CHANGES

Other Books by Carol P. Christ

Diving Deep and Surfacing

Laughter of Aphrodite

Odyssey with the Goddess

Rebirth of the Goddess

Womanspirit Rising
(co-edited with Judith Plaskow)

Weaving the Visions
(co-edited with Judith Plaskow)

She Who Changes

Re-imagining the Divine in the World

Carol P. Christ

SHE WHO CHANGES

First published 2003 by
PALGRAVE MACMILLAN™
175 Fifth Avenue, New York, N.Y. 10010 and
Houndmills, Basingstoke, Hampshire, England RG21 6XS.
Companies and representatives throughout the world.

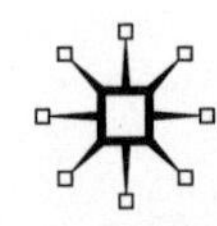

PALGRAVE MACMILLAN IS THE GLOBAL ACADEMIC IMPRINT OF THE PALGRAVE MACMILLAN division of St. Martin's Press, LLC and of Palgrave Macmillan Ltd. Macmillan® is a registered trademark in the United States, United Kingdom and other countries. Palgrave is a registered trademark in the European Union and other countries.

ISBN 1-4039-6083-6 hardback

Library of Congress Cataloging-in-Publication Data
Christ, Carol P.
She who changes : re-imagining the divine in the world / by Carol P. Christ.
p. cm.
Includes bibliographical references and index.
ISBN 1-4039-6083-6
1. Process philosophy. 2. Feminist theology. 3. Femininity of God. 4. Goddess religion. I. Title.

BD372.C48 2003
146'.7—dc21

2003046737

A catalogue record for this book is available from the British Library.

Design by Letra Libre, Inc.

First edition: August 2003
10 9 8 7 6 5 4 3 2 1
Printed in the United States of America.

THIS BOOK IS DEDICATED WITH LOVE TO THE MEMORY OF

CHARLES HARTSHORNE

AND

RACHEL CARSON

AND TO THE LIVING INSPIRATION OF

JOHN B. COBB, JR.

Once this rocky coast beneath me was a plain of sand; then the sea rose and found a new shore line. And again in some shadowy future the surf will have ground these rocks to sand and will have returned the coast to its earlier state. And so in my mind's eye these coastal forms merge and blend in a shifting, kaleidoscopic pattern in which there is not finality, no ultimate and fixed reality—earth becoming fluid as the sea itself.

—Rachel Carson, *The Edge of the Sea*

Contents

Preface

This book is the fruit of a love affair with a philosopher who died in the year 2000 at the age of 103. Two stories suggest why I am drawn to his understanding of the world. In his fifty-fifth year, Charles Hartshorne, a professor of philosophy at the University of Chicago and a lifelong birdwatcher, went back to school and engaged in reading and field research in order to answer the question, "Do birds enjoy singing?"[1] Hartshorne knew that the answer to this question is an unqualified yes. He did not deny that birds sing to attract mates and to control territory as ornithologists allege, but he was certain that birdsong would not be so beautiful and so varied if birds did not also love to sing. This argument is part of several larger ones in his work: that life is meant to be enjoyed (in the widest sense) and that human beings (and God) are not the only beings in the universe with the capacity for joy. I must confess: I fell in love with this man who loved birds so much that he went back to school to prove that they feel joy in living.

In a collection of essays published in the one hundredth year of his life, Charles Hartshorne made only one change in his earlier work. I will let him speak for himself: "I now regret having for so long followed the routine practice of using the male gender in referring to deity; also in taking man as the name of the species. I became profeminist more than seven decades ago [in the 1920s] but began showing this linguistically less than two decades ago. I have tried to purify some of the offending passages."[2] I must confess, I fell in love with this man who cared so much about treating women fairly that he went back to his old essays and corrected them in his ninety-ninth year.

I heard Charles Hartshorne speak only once. I received a single letter from him, written by hand when he was over one hundred years old, clarifying one of his philosophical positions. Thus, I know him primarily through his writing. Hartshorne was a small man with a heart large enough to imagine freedom, love, creativity, and beauty as the guiding principles of all life in the universe, bold enough to argue that these principles apply to other animals and, in some respect, even to the cells and atoms that make up our bodies and the world body.

Having dedicated my career up to this point to the attempt to create a female train of thought,[3] I am quite surprised to find myself waxing eloquent about a male philosopher, white and dead, old enough to have been my grandfather. When I ask myself how this happened, I find myself responding that Charles Hartshorne's understanding of God as the most relational of all relational beings and as the most sympathetic of all sympathetic powers in the universe is amazingly compatible with my feminist longings. If a feminist philosopher is one who assumes that women are (fully equal) human beings and who takes embodiment, relationality, and embeddedness in nature for granted, then Hartshorne is a feminist philosopher. I believe his ideas have much to commend themselves to feminists interested in religious questions.

These brief reflections indicate why I am drawn to the work of the man who was known as America's greatest living metaphysician when he died. What they cannot convey is the joy I have felt, and the opening of my heart to the world that occurred while reading Hartshorne's thoughts about the love of God for the world, "a love divine, all loves excelling."[4]

INTRODUCTION

"She changes everything She touches and everything She touches, changes."[1] This is a line from a chant of the Goddess movement, yet it articulates an understanding of divine and human power shared by many Christian and Jewish feminists, as well as other spiritual feminists,[2] including those who practice or are influenced by non-western or indigenous[3] religions. Change and touch, process, embodiment, and relationship are at the heart of many feminist re-imaginings of God and the world. They are also at the heart of process philosophy. In this book I invite spiritual feminists to think together about how a feminist process paradigm can help us to articulate the radical difference of our visions from those of traditional theologies and to ground our desires to change the world.

In our anthologies *Womanspirit Rising* and *Weaving the Visions,* Judith Plaskow and I showed that feminist theologies and thealogies[4] have a great deal in common, despite and across traditional religious boundaries.[5] In different ways, feminists in religion are challenging images of God as male, disembodied, and separated from the changing world. In different ways, we are seeking spiritual understanding that can inspire us to create a better world for women, for all people, and for all beings in the web of life. I was reminded of these connections when I found myself drawing almost as much on feminist Christian and Jewish sources as on Goddess sources in writing *Rebirth of the Goddess,*[6] my feminist thealogy and ethics.

For many reasons, religious boundaries among feminists in religion seem to have become firmer, not more porous, in the past decades. While much has been gained, a sense of a common purpose has been lost. Though the post-traditional Goddess movement has grown and matured in America, Europe, Australia, and New Zealand, and while increasing numbers of feminists are practicing non-traditional, non-western, and indigenous religions, these developments for the most part have not been considered in Jewish and Christian feminist discussions, especially in North America.[7] I do not attribute this to ill will among feminists, but rather to pressure from those who fear that if women come together across religious boundaries, all hell will break loose.[8] Perhaps it should. There are many connections among spiritual feminists that traditional religious identifications mask. Striking similarities emerge when Christian, Jewish, Goddess, and other feminists create rituals that put women, women's experiences, and the divine feminine at the center; when we talk about the relational, embodied, and embedded self; and when we address global capitalism or ecofeminist concerns. Feminists in religion share a critique of male God-language and traditional understandings of divine power. But we share much more than that. Many who have left Christianity or Judaism behind are beginning to acknowledge that we build on positive elements in biblical religions as we create new understandings of divine power.[9] There are a growing number of spiritual feminists who (with greater and lesser degrees of secrecy) participate in Goddess groups and meditate or practice yoga as well as attending (or even leading services in) church or synagogue. Others read and are influenced by Christian, Jewish, or Goddess feminism, yet consider themselves Buddhist, Hindu, Taoist, or followers of indigenous religions. As modern western feminists struggling to articulate spiritual visions, our interests and our practices have much in common that traditional religious categories mask.[10] Without diminishing the importance of any of the debates that have occupied Christian and Jewish feminists in recent years, I would like to invite my sisters to begin again to think together as feminists in religion about the nature of divine power and its relation to the world.

I believe that process philosophy can provide a fruitful foundation for this dialogue.[11] What exactly is process philosophy?[12] Though some Christian feminists have been re-imagining the divine in the world from a process perspective for several decades, process philosophy has yet to become widely understood. This is not surprising given that process philosophy was formulated in a technical language that makes it difficult to grasp its central insights.[13] As implied by the word from which it takes its name, process philosophy states that all life is in process, changing and developing, growing and dying, and that even the divine power participates in changing life. Human and other beings are not things (substances or essences) situated in empty space, as has often been thought by philosophers and scientists, but are active processes ever in relation and transition. Prior to deconstruction and acknowledging a debt to Buddhism, process philosophy understands that because all beings are always changing, there can be no essential or substantial self or thing. For process philosophy, the self is relational, social, embodied, and embedded in the world. Process philosophy offers a convincing account of creative freedom and its limitations in a thoroughly social world. Process philosophy asserts that feeling, sympathy, relationship, creativity, freedom, and enjoyment are the fundamental threads that unite all beings in the universe, including particles of atoms and the divinity. Process philosophy's divinity has a body, which is the whole world.[14] In process philosophy all beings are connected in the web of life. Process philosophy makes no absolute distinctions between human and other forms of life. These are only a few areas of common vision between process philosophy and feminist theologies and thealogies. In addition, process philosophy affirms evolution and scientific knowledge. It also offers what is to me the only satisfying answer to the problem of evil, arguing that divine power is good and ever present but not omnipotent in the traditional sense, because all creation is co-creation. Process philosophy offers a vision that can inspire us to heal the divisions between people and between humanity and nature that rend our world.

Process philosophy is generally associated with the British philosopher Alfred North Whitehead (1861–1947) and the American philosopher Charles Hartshorne (1897–2000). In this book I will focus on the work of Charles Hartshorne, who developed a comprehensive process philosophy of religion as a response to what he viewed as the shortcomings of the understanding of the relationship of God and the world in western theological and philosophical traditions. Since Hartshorne's critique of traditional understandings of God is a response to western views of God and the world, this book will be helpful to those who wish to transform western traditions, especially Christianity but also Judaism. Those in the Goddess movement and those inspired by non-western and indigenous religions will find process philosophy valuable in differentiating our religious visions from traditional western ones. As Hartshorne's view and my own might be called inclusive monotheism,[15] this book will appeal to spiritual feminists whose world-views might in the broadest sense be defined as theistic. However, the process view of divinity, "panentheism," has very little to do with the God of traditional theism, and in fact incorporates many of the insights of traditions called polytheistic, pantheistic, animistic, mystical, and non-theistic. Thus I have addressed this book to all spiritual feminists who have grown up in the ambience of traditional western religions. I believe that the process view can be helpful to all of us as we seek to articulate the insights embedded in our feminist imaginings. Process philosophy offers a radically different way of conceptualizing divine power and its relation to the world that can illuminate feminist concerns and help us to express them more clearly.

Early on in the feminist theological movement it was asserted that feminism and process philosophy were connected. In her groundbreaking book *Beyond God the Father,* Mary Daly claimed Whitehead as a resource for feminist theology. Conferences were held; papers were read; dissertations and books were written. Feminist thinkers as different as Valerie Saiving, Sheila Greeve Davaney, Penelope Washbourn, Marjorie Suchocki, Rita Nakashima Brock, Catherine Keller, Sallie McFague, and Nancy Howell have made insightful and important use of process thinking in their work.[16] But the combination of feminism and process philosophy has not yet jelled as a "feminist process paradigm" that might function as an alternative or complement to liberationist, deconstructionist, postcolonial, and other models in feminist thinking.

The proposal that feminists in religion need to think philosophically might be welcome to some, yet sound strange to others. When feminists began to think about religion, we were feeling our way out of traditional ways of thinking by criticizing traditions, searching for our history, naming our experiences, experimenting with symbols, creating rituals. Perhaps feminist work in religion needed to grow out of experience and ritual practice in a variety of settings before we could think together about philosophical implications and underpinnings. Now the time may be the right time for sustained and shared philosophical reflection by feminists in religion.[17] Its fruits, I imagine and hope, will be a clearer sense of who we are and how we understand the world, a stronger bond of connection as feminists across religious boundaries, and a renewed understanding of the responsibilities we must embrace if life on earth in anything like the form that we know it is to be saved.

While some feminists have considered philosophical understanding central to their spiritual quests, there are many for whom experience has been sufficient. Indeed, some would argue that feminist thealogies and theologies do not need to be grounded in philosophy at all. "Christian feminist thealogies are rooted in women's individual and communal experiences of certain biblical themes and metaphors and in the struggle for justice," it might be asserted,[18] and this is enough. "Jewish feminists," it might be argued, "can focus on reforming law and liturgy without appealing to theology or philosophy."[19] Goddess religion, it has been said, is a religion of poetry, not theology or philosophy,"[20] and it is based in intuition, experience, and community, or in ritual traditions handed down from the past. "Buddhism, including feminist Buddhism, is a religion of meditation, not theology," it might be claimed.[21] For westerners intrigued with Hinduism, practices such as yoga and meditation, and the plethora of symbols, including those of Goddesses, might seem sufficient. "Indigenous religions are experiential and communal, not philosophical," some might say.

Many perceive philosophy as an elite discipline, created by a small number of well-educated and sheltered white men, who have had little experience of or concern for the realities of everyday life, certainly not for the lives of women and non-elite men. If one considers the history of western and non-western philosophical traditions, this assessment is not entirely wrong. Many, if not most, philosophers speak to each other, not to people with whom they share the world, and certainly not to the mothers, wives, sisters, and servants who make their life of the mind possible. One could ask, for example, what relevance the insight of Descartes' "I think, therefore, I am" could possibly have to women who, according to many philosophers, do not even have the same rational capacities as men. Often the insights of so-called great philosophers are rooted in unrecognized gender, class, racial, imperialist, and other interests. If this is all philosophy is or could be, then feminists and others ought to reject it.

Yet at its most basic and fundamental level, philosophy, like theology, is reflection on the meaning of life.[22] Who are we? Why are we here? What is the meaning of a life that ends in death? How are we related to each other and the web of life? How should we live? If philosophy is one of the ways we make sense of our lives, then we are all philosophers. If we are all philosophers, we also all have the capacity to reflect more deeply upon our experiences and assumptions. Socrates understood this when he entered into a philosophical dialogue with a slave who, according to the power relations of his time, would have been considered incapable of philosophical reflection.

I was lucky enough to have been introduced to philosophy by an undergraduate teacher who began his class by stating, "Philosophy is not about what men thought, but about where the truth of things lies."[23] This teacher's words are anachronistic in this deconstructionist age, when we no longer believe so simply in philosophy's power to unveil "the truth." At the time, however, they were an invitation to me—an undergraduate student and a woman—to begin to do my own thinking about the truth or truths of life.

If thinking about the meaning of life is something we all do, then what can more disciplined philosophical reflection add to feminist thinking? My answer is that philosophical conceptions are involved in all feminist constructions and reconstructions of religion. As I will show, we are implicitly making philosophical judgments or assuming philosophical positions all the time. Since we are doing so anyway, I believe it is better that we do so consciously. Otherwise there is a great danger that we will revert to traditional ways of thinking about divine power and its relation to the world without even being aware that we are doing so. Unless we have alternative conceptions clearly in mind, it is easy to fall back on habits of thought that are embedded in the ordinary language we use when we talk about divinity. When this happens, the radical edge of our thinking is lost.

The languages of prayer and ritual, both traditional and nontraditional, are rooted in implicit philosophical assumptions. For example, it is clear to many that biblical language referring to God as Lord and King reflects imperialistic, feudalistic, and hierarchical world-views that are incompatible with the egalitarian visions that inspire most feminists. When Jewish feminist Marcia Falk began rewriting the language of traditional Jewish prayer, she considered substituting Lady and Queen for the well-known references to God as Lord and King. She rejected this solution because it left the hierarchical power relations inherent in kingdoms intact. In some of the earliest prayers she wrote, Falk referred to the divine power as the "Source of Life."[24] In Hebrew, the word "source" is technically feminine, but the metaphor of the Source of Life can be understood as genderless or inclusive. Source in Hebrew and to a lesser extent in English evokes a spring or a fountain, water springing up from under the earth, and the waters of life (including the waters of birth) more generally. In her other prayers, Falk abandoned even this metaphor because it seemed to create distance between the divine power and ourselves, between the divine power and the world. In the Morning Blessing, Falk wrote:

The breath of my life
will bless,

the cells of my being
sing

in gratitude,
reawakening.[25]

In Hebrew the word "breath" is resonant, recalling the "breath of God" that hovered over the waters in the Genesis creation story. In Falk's prayer this resonance is there, but the object of the blessing (if there is one) is left unnamed. In this prayer Falk seems to be saying (following theologian Paul Tillich perhaps) that the divine power is not a being among beings.[26] Without debating the rightness or wrongness of the philosophical conceptions that underlie Falk's prayers, I wish for the moment simply to call attention to the fact that philosophical conceptions inevitably underlie and inform prayers and liturgies.

In Minneapolis in 1993 Christian feminists at the Re-Imagining Conference invoked the divine power, saying:

Sophia-God, Creator-God,
let your milk and honey pour out,
showering us with your nourishment.

Our sweet Sophia, we are women in your image:

With nectar between our thighs, we invite a lover, we birth a child;
With our warm body fluids, we remind the world of its pleasures and sensations."[27]

Widely reported in the press, this prayer caused an uproar in some of the Protestant denominations that had sponsored this gathering of Christian women in celebration of the Ecumenical Decade of Churches in Solidarity with Women declared by the World Council of Churches. While the women at the conference found the Sophia prayer an appropriate way to re-imagine divine power, traditionalists countered that they had gone too far, that the image of Sophia they had created was no longer Christian. The traditionalists might simply have been reacting to the naming of God as female. Yet their objections may also have been philosophical and theological. They may have felt that the physicality of the language in the Sophia prayer threatened long-standing notions of divine transcendence. Again we see that feminist experimentation with liturgy raises questions about how we understand the nature of divine power.

Philosophical issues are also involved in the creation of Goddess rituals and prayers. Should Goddess worshippers invoke Goddess as Lady and Queen? Some of the most beautiful ancient and contemporary Goddess prayers use these images.[28] Should all such hierarchical language be rewritten? What about personal language? Is there a Great Goddess who listens to and answers our prayers? Should we address them to her, to the spirits of all living things, to the wind, the rain, the sun, and each other? Or should we, like Falk in the Morning Blessing, leave the object of our prayers indefinite? Again, my point here is not to answer these questions but simply to point out that philosophical questions and decisions are also involved in the writing and rewriting of Goddess prayers and rituals.

For feminists in other theistic or polytheistic traditions, Goddesses already exist, such as Aphrodite, Isis, Kwan Yin, Kali, Oshun, Changing Woman, and many others. Western women of all colors have been nourished by them. Yet many of these images have been handed down in traditions shaped to a greater or lesser extent by men. As western feminists begin to experiment with non-western religious symbols, we must also ask to what extent symbols that have offered new and exciting possibilities of spirituality for western women have been molded by patriarchal perspectives and concerns. For example, the anger of the fierce Kali has been inspiring to many western women, enabling us to feel comfortable expressing anger at injustices in our own lives and in the world.[29] But Kali is not only fierce and angry. She approaches the world with a warrior's sword and goes into battle against alleged demons. While the anger of the Goddess and her connection to death may have deep roots in the Neolithic past of India, the image of Kali as a warrior is likely to have been developed in relation to Indo-European warrior traditions that no one could call feminist. In adopting Kali as a "feminist Goddess," are western feminists perpetuating elements of militaristic and patriarchal Hinduism without realizing it? Would it be better for western feminists to adopt the compassionate image of Kwan Yin as "She Who Hears the Cries of the World"?[30] Again, questions about the nature of divine power arise out of feminist spiritual practice.

Philosophical assumptions underlie everyday statements we make about the relation of Goddess or God to our daily lives. For example, when suffering occurs in our personal lives, we almost automatically ask, "Why is this happening to me?" "There must be a reason," we answer, and we start seeking the "lesson" we "needed to learn." "The divine power," we say, "would not harm us needlessly or give us more than we can bear. Therefore," we conclude, "all suffering occurs for a reason, as a punishment for something we have done, or to teach us something," for example, how to live less stressfully, or how to be more compassionate or more accepting. This kind of thinking can be found among Christian and Jewish and other feminists, and it is common in western understandings of Hinduism and in the Goddess and New Age movements.[31] Few of us stop to think that the idea that all suffering occurs for some reason is based upon the notion that a single divine power is in control of everything that happens in the universe. At this point, I will neither criticize nor offer alternatives to this way of thinking, but only note that it assumes a very definite understanding of the nature of divine power and its relation to the world.

Tarot cards, I Ching, runes, Goddess amulets, cowry shells, and other forms of divination have become popular among spiritual feminists. For many they are an essential part of Goddess religion.[32] Christian, Jewish, and other feminists who have been influenced by the women's spirituality movement also experiment with divination. Many view divination as activating intuitive or non-rational ways of knowing. This can be useful as well as entertaining and enjoyable. Yet many western women, myself included, have sometimes understood divination as a way of getting information about the future. When divination is used in this second sense, it makes philosophical assumptions about the way the future is constructed. One assumption that might be being made is that there is some perspective (the divine perspective or the perspective of a shaman or seer) from which the future is already determined, already known, or has already happened. How this understanding fits together with human freedom—including the freedom we and others have in the situations we are asking about—is usually not considered. Once again, unexpressed philosophical questions and assumptions underlie everyday experiences of feminist spirituality.

The common thread in all of these examples is that feminist spiritual practice raises philosophical questions about the nature of divine power and its relation to our lives. Feminist theology and thealogy began as radical challenges to traditional ways of thinking about God and the world. It is my conviction that the time is right for spiritual feminists to begin to reflect together and across religious boundaries on what we mean and do not mean as we reimagine divine power and its relation to the world. It is time for us to think philosophically in a more explicit ways. If we do not, we may forfeit the opportunity to think as boldly and as creatively as we might.

In this deconstructionist age, philosophical reflection on religious questions is not popular. Deconstructionists have told us that every attempt to find "ultimate truth" is rooted in a will to power.[33] Those who claim to know the truth, deconstructionists say, have most often used it as a club to bully others into submission. Considering the history of power relations in the world, this judgment is not entirely false. As recently as a few hundred years ago, the "enlightened" thinkers who wrote the American Constitution denied the rights of citizenship to all women and did not consider slavery to be contrary to the ideals of freedom they propounded.

If power relations are embedded in every truth claim, does this mean that there can be no discussion of ultimate questions? Some deconstructionists take this view.[34] I agree with them that we must always be suspicious of the will to power in ourselves and others. If any philosopher, dead or alive, claims to have access to a single unified absolute truth, his or her views should immediately be "deconstructed." However, I believe that the purpose of philosophy can be redefined or redirected in light of deconstruction. What if we conceive of philosophy as a conversation in which we discuss the meanings of our lives on this earth? In this conversation, the stuff of life would be taken as primary, reflection as secondary. Yet it would also be understood that "experience," by which I mean our felt perceptions of our bodies and minds as embedded in personal and structural relationships in the world at any moment, gives rise to questions that are not immediately answered within it. It would be recognized that language and the conceptions it encodes shape what we know as "experience" and thus need to be reflected upon and criticized. In this understanding, what has been called deconstruction is a tool, but it does not have the final word. Experience is not unproblematic, but neither can it be reduced to language. The body, others, societies, and the whole physical world stand in a complex relation to language, sometimes being interpreted through its lens and at other times challenging the adequacy of the conceptions expressed or implicit in language.

This book is a feminist philosophy of religion. How does philosophy of religion differ from theology? Religion has two aspects, which can be called participation and reflection. On the one hand, we express our sense of meaning of life when we participate in symbols systems and communities. But we also reflect on the meaning of our symbols and the nature and intentions of our communities. When this reflection arises out of and remains with a particular community, we call it theology or thealogy.[35] While both theology and philosophy make statements about the nature of reality, theology's statements are based in the histories and symbol systems of particular communities.[36] When reflection is generalized so that it is no longer tied to a particular community, we call it philosophy of religion. Philosophy of religion speaks or attempts to speak of meanings and values that are fundamental and basic in life.[37] As we are finite and limited, our understandings will inevitably be influenced by our perspectives, and thus understandings of fundamental values will always be approximate.[38] Therefore, all philosophical statements are implicitly qualified by phrases like "it seems to me" or "as best as I can understand it." Many if not most philosophers have not recognized this.

For me, philosophical reflection differs from conversation in relative, not absolute ways. Philosophy pays more attention than most conversation to consistency, the relation of ideas about one thing to ideas about other things, and to coherence, the ways in which a set of ideas come together to form a world-view. Though sometimes ignored in ordinary conversations about meaning, questions of consistency and coherence are quite important when one of philosophy's tasks is envisioned as creating or articulating new or non-traditional world-views. If philosophy is a conversation, we would assume, as we do in the best conversations, that no one person has the whole truth and that no one has the right to impose her views on the others. Taking the lesson of deconstruction seriously, we would understand that all philosophical systems will always be partial and limited because there is no direct access to "the truth," a subject I discuss in chapter six. But this does not mean that all ideas about the world are equally valid. We can ask if ideas help us to understand ourselves and the world as we know it better. We can ask if our experiences of Goddess or God seem to reveal a divine presence in our lives.[39] We can ask how one idea fits together with other ideas to shape a world-view. Philosophical reflection can also help us to become conscious of the potential or actual political and moral consequences of ideas. We can ask how the views of the world we have constructed influence individual, communal, and global moral decision-making. In this sense philosophical reflection is one of the tools we have for understanding and changing ourselves and the world.

There are many commonalties of interests between feminist work in religion and process philosophy. However, there are several impediments that inhibit making this connection, and these need to be recognized at the outset.[40] Like many philosophies developed by men, process philosophy began as a conversation among philosophers, living and dead. The leading process philosophers, Alfred North Whitehead and Charles Hartshorne, often addressed themselves primarily to other philosophers, assuming extensive knowledge of the history of philosophy, which makes it difficult for the intelligent general reader to pick up their books and read them. Luckily, both also addressed several shorter books to general audiences.

In order to read Whitehead's most profound ideas about the nature of God, one must wade through his opaque and difficult *Process and Reality*, a work filled with neologisms (made-up words and words used in other than their normal sense) that is daunting even to professional philosophers.[41] Whitehead believed that philosophy should be modeled on mathematics,[42] which means that his goal was to use language with a kind of numerical precision, not to communicate with anyone unfamiliar with his technical vocabulary. Unlike Whitehead, I believe that philosophical ideas can and should be expressed in ways that can be understood by a wider audience.

In part because of his extraordinarily long life and also because he learned something from ordinary language philosophers, Hartshorne was able to disentangle many of his central ideas from the technical language of philosophical tradition and to make them accessible. Hartshorne's *Omnipotence and Other Theological Mistakes,* published when he was eighty-seven years of age, is personal, embodied, witty, and deeply concerned with saving the world, yet it contains his major ideas on religion. The same can be said of the collections of his essays, *Wisdom as Moderation* and *The Zero Fallacy,* published when he was respectively ninety and one hundred years old.[43]

The reader who turns to Whitehead's *Religion in the Making* (which is addressed to a general audience) will find a set of lectures on the evolution of religious consciousness using notions of religious progression that seem outdated and ethnocentric. Whitehead divides religions into higher and more primitive forms: beginning with primitive religions based in ritual and emotion, moving on to (more developed) tribal religions based in myth and belief, and ending with (superior) individualistic religions based in rational reflection.[44] Feminists and others interested in embodiment, embeddedness in nature, and community will most likely find more to value in ritualistic and tribal forms of religion than Whitehead did, and correspondingly we may be more aware of the limitations of individualistic rational religion. Whitehead also expressed the view (common in his time) that despite the waxing and waning of civilizations, the course of human history can be thought of in terms of spiritual advance.[45] There is evidence that Whitehead began to question this view in light of Hitler.[46] Evolutionary optimism is less and less plausible when we consider the widespread warfare, suffering, and environmental destruction characteristic of the last half of the twentieth century. Hartshorne lived throughout the whole of the twentieth century, and optimism about the fate of the earth became increasingly untenable for him. Thus, there is an immediacy of ethical concern in Hartshorne's later works not found as clearly in Whitehead's.

Whitehead's ideas have been made more accessible by his followers in the field of religion, including John Cobb, David Ray Griffin, and Marjorie Suchocki.[47] Yet they have placed his ideas in a Christian context, often emphasizing an incarnational Christology, which makes their work of less interest to those seeking a more general philosophy of life and religion. Hartshorne combined some of Whitehead's ideas with his own, developing a full-fledged philosophy of religion, yet his work has received less attention than Whitehead's. It is time to redress that balance. In this book I will focus on process philosophy as articulated by Charles Hartshorne. There are a number of good reasons for this. Two have just been mentioned. One is that his later works are accessible to the general reader, which means that readers of this book can turn to (at least some of) Hartshorne's writings with relative ease. The other is that his philosophy of religion is fully developed and not addressed to a specifically Christian audience.

Unlike many philosophers of his time (and ours), Whitehead did not believe that women are inferior; indeed he argued publicly for women's right to vote.[48] Hartshorne considered himself "profeminist"[49] and was a vociferous advocate of women's right to abortion.[50] Hartshorne's "tyrant God" has much in common with feminist theology's "patriarchal God," as Hartshorne himself recognized. "The feminists' complaint that they have been asked to worship a male deity seems pertinent and well founded," he said. "'Men are masters' easily fits the tyrant conception of God, whose function it is to command creatures merely to obey." Hartshorne also understood the appropriateness of female imagery for divine power. He said that "the idea of a mother, influencing, but sympathetic to and hence influenced by, her child and delighting in its growing creativity and freedom," is more compatible with the process understanding of divinity than traditional conceptions of male power.[51] He even went so far as to say that "femininity is closer than masculinity to the truly eternal and fundamental, or to the divine. . . . Much Western theology . . . has been viciously over-masculine."[52] He also stated that one of the reasons he could not accept any theories of divine incarnation in the person of Jesus of Nazareth was that "any such theory strongly suggests the idea of deity as highly spiritualized masculinity."[53] I read these comments, which were never systematically developed, as expressing Hartshorne's agreement with what he knew of the feminist critique of God. Hartshorne's God is fully relational and embodied in the world, compassionate and loving, concerned to promote freedom and creativity in human beings and all beings in the web of life. Such a vision challenges conventional hierarchical masculinist understandings of divine power, resonating with feminist imaginings.

In his later works, in sympathy with feminists and to express his own conviction that God is not male, Hartshorne sometimes referred to God as "He-She" and "Him-Her."[54] This was quite an unusual step for a male philosopher of religion. Hartshorne's decision to use dual pronouns—unlike the more common avoidance of masculine pronouns—makes the point that God is appropriately imagined as female. From here on I will refer to the divine power intended by Hartshorne and process thinking as "Goddess/God." I use "Goddess/God" in order to emphasize that divinity is definitely not to be construed as exclusively male. Though Hartshorne never to my knowledge considered using the word "Goddess," I find "Goddess/God" consistent with his conception of divine power and with his increasing sensitivity to the power of genderized language. It is my conviction, possibly not shared by Hartshorne, that the word "God" connotes the male divinity of biblical religion for most westerners.[55] Thus I find it necessary and appropriate to say "Goddess/God." I am perfectly comfortable using "Goddess/God" in philosophical discussion because I understand the divine power to be beyond gender or inclusive of all genders. However, I also believe that the word "Goddess" must be spoken along with "Sophia," "Shekhina," and other female names in prayer and liturgy if we hope to break the hold of "God" as masculine and male on the human mind. I will continue to refer to the divinity conceptualized in traditional western philosophical, theological, and religious contexts as "God," because I believe that "He" is most often and most appropriately understood as male. I will not change direct quotations from Hartshorne or any other writer.

My perspective has been deeply influenced by Hartshorne, and I am proud to acknowledge that and to call myself a process thinker. However, I was moving in the direction of process philosophy even before I knew it existed. When he was referred to as a Whiteheadian, Hartshorne insisted that he had come to many of the insights known to the world as Whiteheadian on his own.[56] He also noted that while Whitehead had focused on the relation of process thinking to modern science, he had focused on philosophy of religion. Just as Hartshorne had different interests from Whitehead, so I have different interests from Hartshorne. This difference is based in history (I was born half a century after Hartshorne), in class (Hartshorne came from an educated family in which philosophical ideas were freely discussed, while I did not come from an intellectual background), and in gender.

Both Whitehead and Hartshorne understood process thinking to be part of a conversation with philosophers living and dead, and they continually situated their ideas in relation to those of Plato, Kant, Leibniz, and others. Clearly they both felt themselves to be very much a part of the white male western philosophical tradition. Despite having had the privilege of an elite university education (on scholarships gained through a combination of good luck and hard work), I never felt at home in the philosophical traditions of the west, much as I struggled to place myself within them. This means that the relationship of process thinking to the history of western philosophy is less interesting to me than it was to Hartshorne and Whitehead and that some technical distinctions in process philosophy that were created to solve problems created by other philosophers are not crucial for me. Because Whitehead viewed philosophy on the model of mathematical exactitude, he coined new words and used well-known words in specially defined ways in order to express his ideas. It did not particularly bother him that even trained philosophers would have to struggle to understand what he was trying to say. In contrast, I understand philosophy on the model of conversations we have with each other about the meaning of experience and life. I view unnecessarily difficult and technical language as part of a process (established centuries ago) of controlling knowledge and limiting access to it. The result (intended or not) is that large numbers of otherwise intelligent people do not come into contact with ideas that might be useful to them and also that they do not have an opportunity to extend or criticize such ideas. Whitehead and Hartshorne studied in times when higher education was still considered the privilege of a few. Perhaps this fact allowed them to feel comfortable addressing themselves primarily to other philosophers. I entered university at a time when higher education was beginning to be considered the right of every high school graduate and when publicly funded scholarships and low-cost state universities made this a possibility.[57] I also had the good luck to begin writing at a time when large numbers of recently educated women were hungering for new ways of understanding spirituality. Knowing that my work would be read by many intelligent women and men who were not part of the academic community, as well as by members of the guild, meant that I would work to express difficult ideas in the simplest possible fashion and avoid technical jargon as much as possible.

Though Hartshorne considered himself a feminist, questions concerning women and gender are more important to me than they were to him, for obvious reasons. He never read in the works of allegedly great thinkers that members of his sex were incapable of rational thinking. While Hartshorne's student years were interrupted by World War I, mine were shaped by student protests against racism, poverty, and the Vietnam War, as well as by an emerging feminist movement. While his experiences as a nurse to soldiers wounded in combat influenced Hartshorne to begin to question progress in human life, my experiences in the 1960s and 1970s encouraged me to question all authorities. My position as a woman studying traditions created by men and my later feminist consciousness led me to be far more suspicious of all so-called great thinkers than Hartshorne ever was. Hartshorne never found symbols to replace the biblical religion in which he no longer believed.[58] Like other feminists who are participating in the process of re-imagining, I have found alternative symbols, in my case through the feminist spirituality and Goddess movements. This has influenced my understanding of process philosophy, as will become clear in chapter eight. For these and other reasons, my understanding of process philosophy differs from Hartshorne's.

I began my graduate work in Religious Studies at a time when very few women had been accepted into the field. I was younger, blonder, and taller than almost all of the men with whom I studied. I was often perceived more in terms of my body than my mind. Whether I wanted to or not, I was forced to think about the relation of mind and body and about the mistakes philosophers and theologians made when they wrote about women. For me these were not only intellectual questions but quite truthfully matters of life and death. In the years when I felt most isolated, thoughts of suicide frequently entered my mind. In this context it is not surprising that I was unable to ignore my body in order to fit my mind into the world-views created by men who themselves had attempted to separate their minds from their bodies. Thus, like other feminist process philosophers I insist that process philosophy understand the body and the world as the body of God in physical and material terms—not solely as metaphysical concepts.

Early in my graduate career, I became aware of the feminist movement, and in it I found myself being heard for the first time in years. Learning that other women who did not seem to be crazy felt as I did was profoundly healing. For my doctoral thesis I wrote about Holocaust theology as expressed in the narratives of Holocaust survivor Elie Wiesel.[59] In choosing Wiesel, I decided to focus on a point of crisis in western theology: How could a good, loving, and powerful God have allowed the Jews to suffer and be killed in the concentration camps? I felt that in literature the relation of theological questions to our lives could be addressed more clearly than in the more distanced theology I was studying. As I experienced Christianity's relationship to anti-Judaism and became increasingly aware of the sexism of religious language, I found myself less and less able to participate in Christian worship.[60] While working on my doctoral thesis, I was invited to join the first Conference of Women Theologians at Alverno College in 1971. There, I gained the courage to found the Women's Caucus—Religious Studies affiliated with the American Academy of Religion and the Society for Biblical Literature and to propose the creation of the Women and Religion section in the American Academy of Religion. I was thrust into a kind of limelight at a young age.

During this time, I was nourished in my own spiritual awakening by women writers. I read and discussed the works of Doris Lessing, Ntozake Shange, Adrienne Rich, Denise Levertov, Kate Chopin, and Margaret Atwood with friends and students. These writers named my spiritual understanding at a time when I felt very alone and there was little feminist work in religion to turn to. I used the work of women writers to discover and create a language of women's spiritual quest in my first book, *Diving Deep and Surfacing.*[61]

While I was working on *Diving Deep and Surfacing,* I chanced to meet a young woman named Starhawk who introduced me to the Goddess. Starhawk's words resonated with my longing for a female God, my sense of spiritual connection to nature, and my desire for rituals that expressed my understanding of the divine in the world. My second book, *Laughter of Aphrodite,*[62] documents my intellectual and personal journey from Christianity and Judaism to the Goddess. During the writing of it, I was exploring the body of the Goddess at her temples in Greece during the summers and celebrating the seasons and cycles of sun and moon with a group of women in California.

While finishing *Laughter of Aphrodite,* I was urged by colleagues and editors to write the first Goddess thealogy. I signed a contract and shortly thereafter resigned a full professorship in Women's Studies and moved to Greece. I understood that my decision to leave Christianity for the Goddess had severely limited my academic options. I also sensed that in Greece I would find healing and learn to write about the Goddess in a more embodied way. The book did not come easily, and indeed it was ten years before I could finish it. In the interim I began leading Goddess pilgrimages to Crete for women.[63] On the first pilgrimage I underwent a profound spiritual transformation that brought knowledge of the Goddess more fully into my body. I wrote about this journey in memoir form in my third book, *Odyssey with the Goddess,*[64] and will reflect on it later in this book.

After completing *Odyssey,* I was finally ready to write the Goddess thealogy that was published as my fourth book, *Rebirth of the Goddess.* This was a very difficult book for me to write as I had no models and no tradition of thinking to refer to. As always my good friend and theological companion Judith Plaskow read and responded to what I wrote. Yet I had written in virtual isolation and had not discussed the ideas I was developing with colleagues as I wrote. With some trepidation I sent a draft of the book to two well-known Christian theologians I knew only slightly, John Cobb and Gordon Kaufman. I hoped that they might recognize my book as a work of theology. To my great delight, they both liked it very much, and I found myself being mentored for the first time since my undergraduate days.

John Cobb not only appreciated *Rebirth* but also claimed it as a work of process theology. This surprised me because I had consciously referred to process philosophy on only three pages of a three-hundred-page book. It seemed that I had intuitively and unknowingly worked my way into a process way of thinking as I reflected on the meaning of the re-emergence of the Goddess in contemporary life. After meeting with John Cobb to discuss *Rebirth of the Goddess,* I began studying process philosophy seriously for the first time. I had just started Hartshorne's essay "Do Birds Enjoy Singing"[65] when Cristina Nevans came to visit me at my home in Lesbos one spring. We were entranced by Hartshorne's love for birds and for life and ended up reading his essay out loud to each other in my living room. We began to notice birds singing all around us. As I continued to read Hartshorne, I was astonished at how compatible his thinking was with my own. And so this book was conceived. Throughout the writing of it, Cristina, Judith, and John have read every word, supporting and challenging my understandings. In them (and with the help of e-mail, fax, and telephone) I have found the community of like-minded thinkers I have always longed for.[66] Judith commented that in a sense I had come full circle from dissatisfaction with disembodied academic theology through literature, experience, and ritual to the open, embodied, and dialogical thinking about religious questions I had envisioned sharing when I began graduate study. In the past years, I have often wondered how different my career might have been if I had been introduced to Hartshorne's work earlier or if I had studied with him. Easier? For sure. As interesting? Who knows? I like to imagine that the support for my work that I have received from Hartshorne's former student and friend John Cobb at this time in my career is something like the mentoring I might have received from Hartshorne himself. Life is full of surprises.

Though this book owes a great debt to Hartshorne, it is a new "creative synthesis."[67] Placing ideas in a new and different context inevitably changes them. Hartshorne perhaps intuited that his work might appeal to feminists, but he never developed that intuition. Adding my voice to those who are thinking about feminist spirituality and process philosophy together, I suggest that spiritual feminism can be enhanced by process philosophy and that process philosophy can be enhanced by feminism. It is possible that one of the reasons that process philosophy has not found more adherents is that it departs from widely accepted and often unconscious masculinist habits of thought. For traditional thinkers a rational "man" is one who controls his body, is independent of relationships, and does not let emotions like sympathy interfere with his judgment. In contrast, process philosophy affirms the body-mind continuum, the social or relational self, and sympathy as an essential way of knowing. Process philosophy makes qualities and capacities that have been associated with weakness, vulnerability, and women central. It is not surprising that process philosophy might cause traditional thinkers to feel uncomfortable. Making its criticism of the gender bias of western thought explicit rather than implicit could enable process philosophy to be more fully and clearly understood and might even gain it a wider hearing. It is my hope that process philosophy will appeal to a wider group of feminists in religion if it can be expressed in ways that make its insights accessible. The project of expressing difficult philosophical ideas in relation to ordinary experiences and in ways that can be easily understood is itself a philosophical task with moral and political implications. If process philosophy is correct that there is nothing static or fixed in the universe, then ideas are transformed when the context and conditions of access to them are changed. If more feminists in religion become interested in process philosophy, it will be changed as we add our insights to it. At the same time, process philosophy can help feminists in religion to become clearer about what we mean and do not mean when we speak about the self, relationships, the world, and Goddess or God. A feminist process paradigm will articulate the radical difference between our visions and traditional western understandings of God and the world. A feminist process philosophy can help us to understand why the fate of the earth is to a large extent in our hands. A feminist process philosophy can help us to understand what we can do to change the world. "Change is. Touch is. Everything we touch can change."[68]

Chapter One

Problems with God

He is an old white man with a long white beard, dressed in blue, white, or lavender robes, sitting on a golden throne in heaven, surrounded by clouds. He created the world out of nothing. He rules it with His laws and could wipe it out at a moment's notice, if He chose. At His feet is a heavenly host of angels in white robes, with harps. (Once the harps were swords and the heavenly hosts were the army of God, defending His heavenly palace.[1]*) When we die we will go to heaven—if we are good—to live for all eternity with God. God loves the world and its creatures. But He sometimes gets angry and unleashes His wrath on the sinners. His punishment is always just. At the last judgment, He will separate the wheat from the chaff. If we do not follow His will, we will be punished by being sent to hell to be burned in eternal flames, along with Satan. Satan is also a man, naked because of his sin, and he has a forked tail. We must be very careful, or we will end up "down there" with him.*

For many of us in western cultures, feminist criticism of religion began with a protest against this familiar image of God as an Old White Man found in traditional piety. This God is known through the images of Lord, King, and Father. Each of these images is exclusively masculine, and feminists argue that this creates the impression that the highest power in the universe is male. The masculinity of God makes it difficult for women to see ourselves as being "in the image of God" and to affirm our own power. The pronouns used to refer to God—He, Him, His, never She, Her, Hers, nor even It or Its—reinforce the message. It does not matter, feminists in religion assert, that Sunday school teachers and theologians have told us that "God is not really a man." In the case of our language for God, as Mary Daly said, "the medium is the message."[2] Despite protestations to the contrary, the language of prayer and ritual—assisted in Roman Catholicism and Christian Orthodoxy by painted images, including those of God the Father with a long white beard—creates a picture in our minds that is hard to erase. Whether we find this image comforting or disturbing, most of us are familiar with it.

This picture is based in biblical imagery and elaborated in folklore, literature, liturgy, theology, and art in the history of Christianity and, to a lesser extent, Judaism. Its emphasis on sin and punishment is more Christian than Jewish, but Jews too have hoped for heavenly blessing and feared the wrath of God. This picture has been so widely disseminated in both high art and popular culture, from paintings and literary texts labeled "great" by historians to comic books, graffiti, and cartoons, that it crosses religious boundaries. There are few of us, I would argue, for whom the word "God" does not conjure up this picture, if only fleetingly. This is the God that most atheists and agnostics have rejected as cruel and sadistic, as no longer required by modern science, or as too distant from the world to have any real effect upon it. This is the God that many Christians and some Jews hope will reward them in heaven, yet fear may punish them for their sins. If this picture of God is entertained by as many believers and non-believers as I think it is, then it is also part of what western spiritual feminists are up against when we begin to re-imagine divine power.

There are a number of conceptions embedded in this picture. The image is hierarchical, based in ancient and feudal notions of kingship—God is above the world and rules it like a king. It is patriarchal, based in notions of the father's power as supreme and unchallenged. It is racist—the heavenly Father is white, while sin and evil are dark and black. It is dualistic, separating the world into above and below, higher and lower, earth and heaven, time and eternity, good and evil. God's realm, heaven and eternity, is both higher and better than earthly reality. If we are good, we will live forever with God. Finite life is not accepted. Morality is based upon hope of reward and fear of punishment, the assumption being that human beings will do good and avoid evil only because of God's promises and threats. God is understood to be capable of condemning some of his children to eternal damnation. While we are horrified when we learn that some human parents torture or abandon their children, we are taught that God's willingness to abandon some of his children and to allow them to be tortured for all eternity is beyond reproach.

It might be asserted that this is not the way educated people think about God. It is true that educated people, philosophers, and theologians have more complex and nuanced understandings of God. Yet even the "great" poets Dante and Milton did not question the basic outlines of this picture; rather they elaborated and embellished it. Nor did Michelangelo criticize it when, continuing a long iconographic tradition, he painted God as creator of the universe with a long white beard and flowing robes on the ceiling of the Sistine Chapel, and Christ as judge of the living and the dead on one of its walls. Even though some theologians have modified it and even though some Sunday school teachers have told us it is not true, the traditional picture of God continues to be reinforced when the Bible is read in a sacred context, in traditional prayers, and in traditional Christian hymns.

Part of the reason this picture of God is not questioned is that theological and philosophical traditions have subtly and not so subtly reinforced it. Theologians and philosophers have for the most part not questioned the hierarchical notions of power inherent the images of God as a Lord, a King, or a patriarchal Father.

Quite the contrary, they have provided philosophical justification for understanding divine power as unlimited. In "classical theism," so named because its ideas of perfection are taken from "classical" philosophy (especially from Plato, Aristotle, and the Neoplatonists), perfect power is imagined as absolute and unlimited power.[3] That which is perfect must also be unchanging because to change is to become more or less than perfect. That which is unchanging also cannot be related to anything else, because to be related is to change and to change is to be imperfect. Therefore, God cannot be related (in any way that would limit God's power) to the world or its creatures. Therefore, as Hartshorne put it succinctly, God's power is that of a tyrant who is not required to take any account of the needs of his subjects.

Most Christian and many Jewish theologies can be viewed as attempting to interpret God of the Bible as Lord, King, and Father through the lens of classical understandings of power and perfection. Hartshorne names the following theologians as among those exemplifying this trend: Philo Judaeus, a Jewish theologian of the first century CE, early Christian theologians Augustine and Anselm, Islamic theologian al-Ghazali (1058–1111 CE), medieval Christian theologian Thomas Aquinas, Protestant reformers Luther and Calvin, and, more recently, the philosophers Descartes, Leibniz, Kant (in his *Ethics*), and the American theologian Jonathan Edwards. He adds, "the list could be very long."[4] Indeed elements of this kind of thinking are found most theologies. This does not mean that theologians in the classical tradition do not also speak of God as love. However, for them the love of God is more difficult to explain than his power, and God's love for the world is always subordinated to his awesome and unlimited power.

The problems found in classical theism are summed up as "six theological mistakes" by Hartshorne. I will discuss them in a moment, but first I wish to share the story of how I encountered them, because I am certain that elements of my story are shared by others who have studied theology or asked theological questions. Having been brought up to think of God as above all loving, I was mystified when I discovered that most of the theologians I was asked to read, as well my professors and the other students (all of them male), seemed to assume that God was absolutely separate from the world, absolutely powerful, and absolutely untouched by relationships. For them God's judgment was just and not tinged with compassion. They dismissed Saint Teresa of Avila's assertion that a loving God could not and would not sentence any soul to eternal damnation (which made sense to me) as sentimental, wrong-headed, and probably heretical. For my professors and the other students, the self (both human and divine) was conceived as independent in such a way that it was difficult to imagine how it could be related to others. One night one of the students in my program read us a paper entitled "A New Proof for the Existence of Other Minds," while his wife served cake and coffee. Only Judith Plaskow and I found it silly that this man could find it necessary to "prove" the existence of the mind of the wife who made his academic life possible and the minds of those of us who were at that moment sitting in his living room listening to him read.

My professors and most of the other students also thought that mystical or spiritual experience could have nothing to do with the senses, the body, or nature. As an undergraduate, I had written my senior thesis on "Nature Imagery in the Prophets." I had imagined that when the prophets said that the mountains "rejoiced" and the trees of the fields "clapped their hands" on the day of redemption, they meant that all of nature is included in God's salvation and that other than human forms of life have some degree of feeling. While being examined on what I had written, I was asked if the prophets' image of trees clapping their hands on the day of redemption should be understood as "an aesthetic flourish." I did not understand the question. Later in graduate school I learned that for most thinkers in the West, redemption involves God and humanity, but not nature. I also learned that the attribution of feelings to trees is called the "pathetic fallacy," defined as the error of attributing feelings to non-human beings. Though I did not agree, I was beginning to understand why my examiner could not imagine trees rejoicing on the day of redemption.

To my great surprise and consternation, I also discovered that most of the "great" theologians past and present, all of them male, assumed that women were inferior to men, with a lesser quotient of mind (which was in the image of God) and a higher quotient of body (which was not). Though they might not have stated it so directly, the faculty and male students saw no reason to disagree with this notion. Moreover, they all seemed to pay a great deal of attention to my body, but very little to my mind. When I finally had the courage to write on Karl Barth's view that woman will always be second and subordinate to man's initiative, precedence, and authority, the professor tossed my paper aside, saying that this never seemed to him to have been an important issue. And when Judith Plaskow and I proposed doing an examination in historical theology on the history of ideas about women, our professor pounded his fist on the table and bellowed: "Not for me, you're not! Women are not a doctrine."[5]

The problem of evil also troubled me. I understood that soldiers were killing children and old people in my name in Vietnam. I wore a pin with a starving Nigerian child on it to remind me of famine in Africa. After reading Holocaust survivor Elie Wiesel's *The Gates of the Forest,* I became obsessed with the question of how a just and loving God, who had the power to stop the Nazis from burning Jews in the gas chambers, did not do so.[6] When I tried to broach this topic with one of my professors, he waved his hand in dismissal of my question, asserting that for Christians that problem had already been solved by the cross of Christ. My questions about Vietnam, Nigeria, and the Holocaust were not heard.

I knew that there was something wrong with the theology I was studying, but it took me a long time to articulate my objections to it clearly and consistently. (Hartshorne would have helped immensely, but his works were not on any list of books I was asked to read, and to be fair, he had not yet written *Omnipotence and*

Other Theological Mistakes.) My first attempt to challenge the theologies I was studying was to claim the Jewish theologian Martin Buber as an ally. In his most famous book, *I and Thou,* Buber argued that the "I-thou" relation is one in which two individuals are joined in a moment of meeting that constitutes them both. His understanding of the I-thou relationship between "God and man," "man and man," and "man and nature," while expressed in genderized language that I later came to reject, provided a way of understanding theological questions quite different from that found in most of the other theologies I was studying. Buber chose to use the intimate or familiar German "du," translated into English as "thou," to express the relationship between "I" and "thou." Buber's God as "Eternal Thou" is clearly relational and intimately related to the world. Human beings too were understood to be so fundamentally relational that the question of the existence of other minds could not arise. For Buber, the other person as "thou" is as real to me as my own "I." Buber also understood that relationships with nature, with a tree, a horse, and a piece of mica, could also be I-thou relationships.[7]

When I tried to argue that Buber represented a more relational and less hierarchical alternative to traditional ways of thinking about God and the world, I was told by my professors that Buber was "only a poet." Since I believed that poets often express truths more powerfully than others, I did not at first understand that calling Buber a poet was a way of dismissing him. Eventually I saw that Buber was being called a poet in the same way that the prophet's view of nature was being called aesthetic. According to one view of poetry (though not one I endorse), poets are expressive and emotional, but their ideas are not to be taken seriously. Buber's assertion that human lives and the life of God were thoroughly relational went against centuries of theological assumptions. If he was viewed as a theologian, his challenge might have had to be considered. Because he wrote in a more evocative style than most theologians, it was easier to discount his ideas by labeling them as emotional excess common to poets. When I tried to argue that Buber was also a theologian and philosopher of religion, I was ignored.

My next approach was to hypothesize that the fact that I was a woman might have something to do with my disagreements with so many great male theological minds, living and dead. Perhaps my understanding of God, humanity, and nature was different because my viewpoint was feminine or female. Perhaps I thought (naively, then) that women are more concerned with relationships and feel closer to nature than men. Much later, I would learn that traditional theology had subordinated women, the body, and nature to men, the mind, and the spirit. This helped me to understand why I as a woman had a difficult time figuring out where I fit into the theologies I was reading. I would soon become critical of dividing the world into masculine and feminine ways of thinking. But I would remain convinced that the fact that most women in the history of the world have been involved in the bearing and rearing of children might make us likely to be sensitive to the importance of the body, nature, and relationships in human and divine life.

During my years of graduate study, I heard the philosopher Charles Hartshorne deliver a lecture that I remember as inspiring. At that time I read his book *The Divine Relativity* in which he argues—contrary to the most commonly accepted views—that God is the most relational and related of all beings.[8] I remember having been very excited to read this confirmation of my own intuitions. However, I was not encouraged to pursue my interest in Hartshorne, and therefore I was not introduced to the full scope of his critique of classical theism. I did not learn that the criticisms that he later summarized as six common theological mistakes of classical theism addressed most of the problems I had been having with the theologians I was studying. Nor did I discover that he had already worked out a philosophy of religion and life that came to many of the same conclusions I would struggle to reach on my own many years later as I wrote *Rebirth of the Goddess*. When process theologian John Cobb pointed out the connections between the thealogy of *Rebirth* and process philosophy, I read Hartshorne again. What I found was a fully developed philosophy that could have helped me to criticize the traditions I was studying. I believe that Hartshorne's criticism of classical theism can help all feminists in religion to become clearer and bolder in our criticisms of traditional views of God.

Hartshorne identifies the six common theological mistakes of classical theism as: God is perfect and therefore unchangeable, omnipotence, omniscience, God's unsympathetic goodness, immortality as a career after death, and revelation as infallible.[9] These theological mistakes derived from classical philosophy are incorporated into the common image of God as an Old White Man sitting on a throne described at the beginning of this chapter. Notions of God's omnipotence and omniscience are presumed to be part of the common picture, as are belief in immortality and infallible revelation. Indeed, these ideas are widely accepted and uncriticized. However, the idea of God's unsympathetic goodness, the idea that God is not truly related to the world or affected by it, will probably seem foreign, for God is traditionally understood to care about the world. The idea that he does not would be dismissed out of hand by most Jews and Christians. Incredibly, though, theologians past and present have argued that God cannot really love the world because any form of relationship would inevitably limit his power. "Not possible!" most religious people would say. Yet most of us have no trouble imagining that God can "do anything"—even destroy the world. When we do so, we are conceding that it is more important to affirm that God's power is unlimited than that God loves the world with unfailing love. The notion that God is more powerful than good is so taken for granted by so many people that it can even work its way into feminist conceptions of divinity. Only clarity and vigilance will protect us from unconsciously or unwittingly slipping back into familiar patterns as we re-imagine God. This is why it is important to understand why each of these six theological ideas should be considered mistakes.

God is absolutely perfect and therefore unchangeable. Historically, the roots of this view of God can be found in Plato's ideal of the Good as unchanging. I quote the following passage from *The Symposium* because the ideas expressed in it have influenced the way philosophers and theologians have thought about God for more than two thousand years. "This beauty is first of all eternal, it neither comes into being nor passes away, it neither waxes nor wanes. . . . It [is] absolute, existing alone with itself, unique, eternal, and all other beautiful things partaking of it, yet in such a manner that, while they come into being and pass away, it neither undergoes any increase or diminution nor suffers any change."[10] Here Plato draws a sharp contrast between the time-bound world that we inhabit and the eternal. Change is what separates our world from the eternal. In our world, things come into being or pass away. In our world, things are born, grow, and die. In the phase of growth, things increase or become more than they were; as they begin to die or disintegrate, things become less than they were. That which is perfect cannot change; otherwise it could become more or less than itself, but this was thought to be impossible, as that which is perfect cannot become more perfect or less perfect. Plato asserts that in order to be free from change, the eternal must exist alone with itself, because relationships inevitably involve change and dependence. The highest Good or, as theologians understood Plato, God, therefore must be free of change, and therefore he must exist alone—that is free of relationships that could cause him to become more or less perfect than he already is. For this God there is no change and no touch.

On the face of it, this seems to be an odd view. There is no doubt that it conflicts with biblical and other images of God's care and concern for his creation and of his overflowing love for his people. Yet it makes a certain kind of psychological sense. We all know that relationships make us vulnerable. If we love deeply, we rejoice in the joy of others and suffer with their suffering. If they leave us or die, we feel bereft. When we feel pain, we sometimes even conclude: "Better not to have a relationship than to suffer so much!" Or: "I wish I had never had any children at all." Something like this is the mentality that lies behind the classical view of God. If God is deeply related to the creatures, then this means that God can become (to put it simply) happier and sadder than God was in the beginning. But if God's feelings can change, then it may be possible that they could change radically. What if God decided: "Better off without all those nasty creatures!" and destroyed the world? If God can change, then perhaps God can also die. And then where would we be? In order to protect against these possibilities, philosophers and theologians starting with Plato in classical Greece developed the idea that whatever is perfect, especially God, cannot change. This solved the question of God's reliability. God would always be there and God would always be the same. However, the notion that God is in all respects unchanging introduced other problems as we will see below.

God's unsympathetic goodness. If God cannot change, then how can God be intimately involved with the creation and the creatures, whose very nature it is to change, to come into being and to pass away, as Plato put it? If God cannot be intimately involved with the changing world, then what can it possibly mean to say that God is love? In classical theology, God cannot really be involved with us. In other words, the goodness of God is an "unsympathic" goodness. God is good, but he cannot accurately be said to care for or about his creation, for if he did, he would be affected by the joys and sorrows of the creatures. And if affected, then changed, and if changed then not perfect. Theologians called this the doctrine of God's aseity or impassibility. Hartshorne found this view despicable. It made God a monster who could not love the world. And it led to equally false assumptions about what it means to be made "in the image of God" or human. If God's perfection is a kind of independence of relationship, then our perfection must be to become as independent of relationships as possible. Out of this mentality ascetic movements were born. More modern notions of the self as freer when independent of relationships arise from the same way of thinking. As Hartshorne said in reaction to such views, "Either God really does love all beings, that is, is related to them by a sympathetic union surpassing any human sympathy, or religion seems a vast fraud."[11] I agree with him, and I will consider the implications of divine sympathy in later chapters.

Omnipotence. This is the theory that God is all-powerful. As perfect, God must be able to achieve his will. This means that God can do anything he wants: create war or bring peace, send rain, thunderstorms, or earthquakes, and so on. It also means that whatever happens must be the will of God; otherwise it could not have happened, as he would not have let it happen if it were not part of his plan. In asserting God's omnipotence, theologians and philosophers were using a more-is-better model. God's power must be more like that of a strong individual than a weak one, more like that of a parent than a child, more like that of a ruler than a slave. As theologians and philosophers tried to imagine a power greater than even these, they came up with the idea that God's power must be the power to control everything that happens in the universe. As Hartshorne says, this idea was felt to be so obvious that alternatives were not seriously considered.[12] It was not asked if conceiving God on the model of an all-powerful parent, strong man, or ruler, was not conceiving of him as an all-powerful tyrant!

Aside from the question of whether a tyrant God is worthy of worship, there are two other serious problems with the notion of an omnipotent God. First of all, if God is in control of everything that happens, explaining human freedom (or the freedom of any other beings) becomes a difficult task indeed. For if we are truly free, then God cannot have determined the things that we ourselves determine using our powers of choice. Theologians have worked long and hard to find a way of having it both ways. But many find their theories unconvincing. Either God determines everything or God determines some things and we determine others. But if we determine some things, then not everything that happens can be said to be according to God's will or purpose. However, if we insist that God determines everything, we come up against a second problem, the well-known "problem of evil." Why do the good suffer and the evil go unpunished? In more personal form: Why is this happening to me? In a more general form: How could God have not have intervened to stop the holocaust, the spread of AIDS in Africa, starvation in India, and all the other bad things that are happening in the world? If God is both all-powerful and all-good and bad things happen, then even the bad things must be part of his will. Perhaps we suffer "for our own good" or "for the good of others." Yet this answer raises other questions: Is God a sadist? Does he enjoy watching others suffer? How much suffering will it take before he intervenes? The answer that is usually given at this point is in a sense to give up: God's ways are not our ways. His will is inscrutable. We may not understand it, but God's will is perfect; therefore, whatever happens has a purpose. That this answer is really no answer at all is obvious even to small children, but when they (or we) object, appeal is made to a higher authority. We are made to understand that to pursue the question further is to challenge the authority of priests, ministers, rabbis, the Bible, God himself.

How could an idea that creates such serious problems for human understanding have been held by so many thoughtful philosophers and theologians for so many years? Clearly this idea also must have a certain psychological appeal. When we are suffering greatly, or when we witness the suffering of others, we all want to ask: Why is this happening? Certainly someone must be in control. The idea that some things happen "for no good reason" and that some people suffer "for no good reason at all" is not easy to accept. We want to believe that there is some time when good will be rewarded and evil will be punished. We want to believe that there is some place where suffering will have a meaning. We want to believe that there is some time when all will be made well. This desire is very human.

The desire to make all things well is the hold that the omnipotent God has on the human imagination. For many traditional believers, the kingdom or reign of God is being established in heaven and hell all the time. Others believe that God's redemption will come when the trumpets sound on earth "at the end of time." Many fundamentalist Christians imagine the faithful being taken up to heaven while their empty cars crash into the cars still carrying sinners on highways and freeways. At least some Christian liberation theologians expect that there will be a time when God as liberator will punish the oppressors, exalt the poor, and establish justice on earth. In New Age thinking, for some Goddess feminists, and for others practicing non-western religions, the notion of divine omnipotence may be combined with theories of reincarnation. In this view, everything works out for the best and according to the will of the Goddess in the long run, because suffering in one life will lead to a higher and better incarnation in the next. A popular version of this theory states that each of us chooses the life we will be born into for the lessons it can teach us. Some even assert that those who suffer most are heroic, because they have chosen to learn their lessons in the quickest possible fashion.

Psychologists, atheists, and materialists have called faith in an omnipotent God childish. In the real world bad things happen to good people, the strong trample the weak, and the rich get richer while the poor get poorer. When humankind grows up and accepts reality, they say, it will find that it has outgrown its need for God as well. Others criticize the omnipotent God on moral grounds. There is no other world, they insist, in which the suffering of six million Jews and others in the concentration camps could possibly be justified. Some things are simply too terrible for us to ever imagine that they could have happened according to divine purpose. Many Jews and others have found it impossible to believe in God after Auschwitz.

Yet some would argue that those who reject the omnipotent God all too often throw the baby out with the bath water. Perhaps the hope that a good father or a good king is in control of everything that happens or will come in glory to fix everything up is childish. Perhaps not all suffering has a meaning. But the desire to live in a more just and loving world need not therefore be discounted. What if there are ways of conceiving a just and loving God other than on the model of omnipotence? What if there are ways of imagining divine power that do not make a mockery of human freedom? What if the divine power can achieve its purpose in the world only if we join with it in the project of co-creating a more loving and just world? Process philosophy offers such a view, as we will see later. I believe such is the only viable one for feminist theology and thealogy.

Omniscience. This is the theory that God knows everything. If God is greater than we are, then God's perspective must be infinitely wider than our own. Our knowledge of the world will always be partial and subjective. We cannot know everything, and we cannot put ourselves fully in someone else's shoes. Only a divine power could know the world perfectly. So far so good, as Hartshorne would say. But theologians generally go farther. God not only knows everything that has happened and is happening now but also everything that will happen in the future. Omniscience and omnipotence go together. It makes sense to say that God knows the future only if it is also assumed that God creates it in every detail. In this case, there is again no room for human freedom. On the other hand if there is genuine freedom in the world, then God cannot know the future until human beings and other individuals with a degree of freedom actually do make the choices that we will make. Some try to get around this problem by saying that we are really free, but God stands outside time and therefore can see the past, present, and future simultaneously. According to this view, the history of the world is like a filmstrip or video that can be fast-forwarded or reversed at the will of the divine power. Yet if the divine power can know infallibly what will happen in the future, it is hard to imagine what role if any is played by individual freedom in the creation of the world.

Again, we must ask why so many sensitive and well-trained thinkers have held such problematic views. And again, we may find an answer in human psychology. When things are going relatively well for us, an open future in which new and surprising things can happen is an appealing prospect. We imagine all kinds of good things coming into our lives, new friends, new lovers, travel, adventure. Yet an open future is uncertain. It can also bring death, disease, loss, and endings. All of these possibilities are multiplied infinitely when we think not only of our own future, but the future of the world. Will warfare, starvation, and destruction of the environment continue until abundant life can no longer be sustained on this earth? Will humanity wake up and change in fundamental ways before it is too late? What can we do? What should we do? When we get anxious about the future, it can be comforting to think that there is a perspective from which all of our worries are meaningless, a perspective from which the future is already determined and already known.

Fundamentalist Christians find it comforting to think that that the end of the world has not only already been determined by God, but has already been revealed by God to them in fairly clear detail. Some Orthodox Jews believe the holy land will be restored to the Jews in peace when the Messiah comes. Even liberation Christians may sometimes appeal to the Bible as having promised that God will triumph over the oppressors of the poor, at the end of time. New Age Goddess worshippers or followers of non-western traditions might scoff at such notions, while turning to astrology, tarot cards, or psychics to find out what will happen in the future.

What if the future is genuinely open and unknown? What if it will be created by a combination of all the free wills in the universe, working in cooperation with or in ignorance of or rejection of the divine power? This is the world imagined by process philosophy and I believe the one most compatible with feminist vision.

Immortality as a career after death. This is the view that after death we will live for all eternity in the presence of the immortal God. Immortality as life after death is assumed in the common picture of God in heaven discussed earlier. Christian understandings of the salvation, in both eastern Orthodox and western churches, turn on the notion that while death is the consequence of the sin of Adam and Eve, eternal life is promised through the death and resurrection of Jesus Christ. In Orthodox tradition it is said that "God became man, so that man might become God." This is asserted in clear recognition of the fact that immortality was previously understood to be what set the Gods and Goddesses apart from man. What theologians meant when they said that man would become God was that man—whose very nature it is to change, to be born, to grow, and to die—will come to share in the unchanging and immortal life of God.

For many the promise of eternal life is one of the main attractions of Christianity. It also functions as one answer given to the problem of evil. This life may be a vale of tears, but in the next, all will be restored. For some, the emphasis on the afterlife in Christianity, felt to be contrary to common sense or thought to turn us away from our responsibilities in this world, is reason enough to leave Christianity. For many modern Jews, the question of life after death is not central. For Hartshorne, modern Judaism's lack of concern about life after death is a strength, not a weakness.[13] However, belief in the afterlife was widespread in Judaisms of the past, and many traditional Jews still anticipate it. Some Jews imagined that in the afterlife men would sit at the feet of the great rabbis and study Talmud in the presence of God.[14] Eastern traditions have other views about life after death. Hindus favor reincarnation, traditional Chinese care for their ancestors' graves and memories, some Buddhists are agnostic on the question, while others incorporate reincarnation or imagine nirvana as some sort of heavenly bliss for the enlightened. Many westerners following these traditions believe in some form of reincarnation.

The expectation of life after death has played almost no role in Jewish and Christian feminist theology. It has either been assumed or ignored, or declared a matter that we cannot know.[15] This is a major departure from traditional Christian and Jewish views, and at least one Christian feminist has noted this and called for the restoration of more traditional understandings of life after death.[16] While it might be expected that in Goddess thealogy the finitude of life would be affirmed, this has not always been the case. Starhawk uses a Celtic myth of the afterworld as a "Land of Apples" in death rituals, while stating that this place may not literally exist.[17] Many in the Goddess movement who have been influenced by the New Age or non-western religions believe in reincarnation, imagining that the soul survives death and moves on to other lives. Many years ago I argued that Goddess religion enables us to embrace finitude and death, while in *Rebirth of the Goddess,* I proposed limited and conditional survival of individuals after death, as long as we are kept alive in the memories of others.[18] Hartshorne argues strongly against the doctrine of immortality as a "career after death." He asserts that death is the appropriate ending to life in the body and is convinced that this understanding is completely compatible with belief in a loving and compassionate God. He sees no reason to wish for or hope for a career after death. Hartshorne provokes us to ask whether focus on any belief or hope that there is another time of place in which we will survive the body is compatible with feminist valuing of immanence, relationship, life in the body, and nature.

Revelation as infallible. This error is not restricted to the Roman Catholic doctrine of papal infallibility but includes all attempts to gain certainty based upon revelation. At first it might seem that this mistake is not connected to the others. But it too is a denial of the changing and fragmentary nature of human life. To say that revelation is infallible is to say that it is unchanging by virtue of its direct link to an unchanging God. Revealed truth thus becomes an exception to the partiality and fragmentariness of human knowledge. When revelation is thought to be infallible, it is said that we can know the divine will for ourselves and the world with absolute certainty and often in concrete detail. Knowledge of the divine will is generally said to have been mediated to by an inspired individual (Moses, Jesus, Buddha, Mohammed, Mary

Baker Eddy, Joseph Smith, a Hindu swami, a New Age speaker); or an inspired book or group of texts (Bible, Koran, Talmud, *Science and Heath, Book of Mormon,* Vedic scriptures); or an individual chosen to represent the divine will to a community (Pope, Patriarch, Dalai Lama). When confronted with moral and intellectual questions about life's meaning, people who believe in infallible revelation will turn to an authority outside themselves to find out what to think and what to do. They are thus relieved of the responsibility of making difficult decisions. In addition, they often reach a certainty that makes them feel morally compelled to convert others, to write laws that others must follow, or to impose their visions through crusades or holy war. Hartshorne argues that belief in infallible revelation is incompatible with the fragmentariness of all knowledge that comes through human bodies.

In the feminist spirituality movement, traditional religious authorities that have enshrined women's secondary status are viewed with suspicion, and this is good. Unfortunately the habit of looking for answers and trusting authorities can slip back into the feminist spirituality. Leaders in Christian, Jewish, or Goddess feminism are sometimes set up on pedestals and trusted implicitly. Christian feminists may view liberation theology as divinely revealed. Orthodox Jewish women might find it difficult to challenge laws or traditions viewed as in some sense divinely instituted. In the Goddess movement, there are those who place trust in channeled information or the visions of psychics, or in allegedly ancient Goddess traditions, west or east. Still others believe that the will of Goddess or God has been infallibly revealed in individual mystical experiences. Feminist followers of non-western traditions might reject the authorities of western religions but give up their judgment to an allegedly enlightened teacher or guru, perhaps a female one.

The psychological appeal of infallible revelation is easy to see. Faced with a myriad of choices, human beings sometimes feel overwhelmed or confused. How much easier life sometimes seems when we simply "follow the rules." Yet allegedly infallible traditions have denied full personhood to women and have enshrined slavery, class, and caste. As many thinking individuals know, so-called infallible traditions often ask us to do things that seem profoundly wrong to us. In addition, great suffering has been caused by clashes among groups, each convinced that the revelation given to them is true. This should cause us to be suspicious of all claims of absolute authority. Process philosophy views all claims to authoritative or infallible knowledge as a denial of the fragmentariness of life. I believe feminists in religion must resist the appeal of any spiritual path that asserts that it has all the answers if only we believe, have faith, or submit our will to someone or something. Too much harm has already been done by allegedly infallible authorities.

The six common theological mistakes of classical theism have been combined with traditional language and imagery to create the patriarchal God criticized and rejected by feminists. To date feminists have focused more attention on symbols and language than on concepts. Yet ideas about divine power as unchanging and unsympathetic, as omnipotent and omniscient, and belief in immortality and infallible revelation are part and parcel of the patriarchal view of God. The project of re-imagining God requires re-thinking conceptions of divine power and its relation to the world. In the following chapters, I will consider process philosophy's alternatives to the six traditional theological mistakes. In response to each of the theological mistakes, process philosophy values qualities or capacities that traditional thinking has identified with vulnerability, weakness, and women. Spiritual feminists will recognize many points of connection between process philosophy and our intuitions and yearnings. Like many feminist theologies and thealogies, process philosophy affirms change and embodiment, touch and relationship, power with, not power over, the world as co-created, this life rather than hope for another, and the fragmentariness of all knowledge.

Chapter Two

Change Is

For process philosophy, the whole universe is alive and changing, continually co-creating new possibilities of life. Every living individual is born, grows, and then dies. The world is a web of changing individuals interacting with, affecting, and changing each other. The body is the locus of changing life. Not to be embodied, not to change, is not to be alive. Change occurs from moment to moment in our daily lives as we are acted upon and act, exercising creative freedom. The universe as a whole is changing in a continual process of evolution. The world is filled with free and creative individuals, related to each other. To a greater or lesser degree, all individuals, including human beings, other animals, cells, atoms, and particles of atoms, exercise creative freedom. Goddess/God[1] is fully involved in the changing lives of every individual in the universe and in the evolution of the whole. Creation is co-creation. The whole world or cosmos is the body of Goddess/God. For process philosophy, change, freedom or creativity, and embodiment are interconnected. Everything in the world is in process. Change most definitely is.

The process view stands in sharp contrast to both classical philosophy and early modern science, which thought the world to be made up of things (substances or essences) situated in empty space. Individuals are not "things" but active processes ever changing in relationships.[2] The divine power also changes and enters fully into relationships. Indeed, Goddess/God is the most related of all related beings and the most sympathetic of all sympathetic powers in the universe.

Process philosophy asserts that knowing that things will not stay the same, knowing we will not stay the same from day to day, from moment to moment, is what makes life interesting and worth living. Yet many of us continue to think: "If only I could accomplish this or that—get a new job, finish a project, find a lover, buy a new house—then everything would be fine." The process of living teaches us that this is not true. We may get the job or finish the project, but life continually provides us with new challenges. This is not some trick of fate. If we were not continually changing, we would not be alive. The dead do not change, though the cells in their bodies do.

Many of us have been taught to fear changing life. We have been taught that things must be just so. "Your room must be clean," our mothers said. "You must be good." Even cleanliness in a room is not a state but a process. A house that is too clean and too orderly is one we feel we cannot live in. On the other hand, a house that is too dirty and too chaotic is one we may feel we do not want to live in. For most of us rooms are continually transforming from greater orderliness and cleanliness to greater disorder and dirtiness, and back again. Goodness too is not a state, like sitting still and being quiet. Goodness is one way to approach the changing world. Goodness has more to do with taking account of the changing needs of others, as well as one's own, than with sitting still and being quiet.

As we grow older, many of us still want things to be just so. Each of us must have an interesting job, a loving partner, children, a nice house, and money. "Otherwise we will not be happy," we have been told. When we can't have one (or more) of these things and have it now, many of us get frustrated and depressed. "Life isn't working out as they told me it would," we say. On another level, we know this is childish. When we are living deeply in touch with changing life, we know that change is the nature of life. But old habits do not die easily. One of the reasons they are difficult to shed is that they are reinforced by centuries of thought, which have filtered down in myriad ways into our daily lives.

In traditional western thinking, change is given an exclusively negative evaluation. Plato stated that change involves coming into being (birth) and passing away (death), as well as increase (growth) and diminution (decay). Though earlier religions and some of those contemporary with Plato (as well as some non-western and many indigenous religions) affirmed the intertwined processes of birth, death, and renewal as sacred,[3] Plato focused on the death-and-decay aspects of change, giving short shrift to birth, growth, and renewal. To put it simply, for Plato change equals death and decay. Since the body is the location of death and decay, the human body and all bodies were found lacking. Plato found change so problematic that he imagined divine power existing totally apart from the changing world, as we have seen. God not only did not have a body; he was also separate from all bodies. This is the first theological mistake.

When theologians tried to combine the unchanging God with the biblical accounts of creation, they imagined that creation happened in an instant and that all the creatures including human beings were created simultaneously, corresponding to eternal ideas in the mind of God. Usually it was added that this creation was "ex nihilo," out of nothing.[4] It was assumed that all of the existing species of plants and animals, including the human species, as well as the heavens and the earth, were created in that single instant and in accord with the perfection of divine knowledge. The number of species was finite and, because divinely caused, perfect. No new species could be added and no species would cease to exist.

Creation was understood to be hierarchically ordered—a vast unmoving pyramid with God on the top, "man"[5] below him, women and children below "man," and all the other creatures below them, divided into animal, vegetable, and mineral. It was said that there was an enormous gulf between humanity and the animals. Only humanity was thought to have a rational and sensitive or feeling mind or soul. For many philosophers and theologians, only "man" had a fully rational mind, capable of communing with the eternal ideas in the mind of God. For them, "woman" was said to think through her body and thus to be more like the animals than man. Great cruelty to animals has been one fruit of this view; great cruelty to women, another.

Of course philosophers understood that "man" is a kind of animal. Aristotle called him the "thinking animal." "Man" it was said, had two natures, a rational nature and an animal or bodily nature. These two natures, it was thought, were continually at war with each other. Whereas reason should have been able to rule the body, all too often, it seemed, the body asserted its own needs and desires. The practice of asceticism, in the East as well as the West, arose out of the attempt to control the unruly body through denial and sometimes punishment. While women also practiced asceticism, the literature of asceticism, written primarily by men, is filled with images equating the temptations of the body with women and the female body. Instead of accepting the changing body as part of the self, asceticism attempted to deny it. Great cruelty to the self and the body have all too often been the fruits of this view.[6]

Women have been the scapegoats in much traditional thinking about change and the body. Most classical philosophers and theologians thought that "woman" was more part of the changing world than "man." Hers was the body that grew larger in pregnancy and then smaller again, hers was the body from which milk and menstrual blood flowed, hers was the body that stopped bleeding and became old. She was the one who cared for children and tended the sick and the aging. His was the mind that contemplated the unchanging God. He did not change, but she certainly did. Thus, it was thought, "man" was a more perfect creature than "woman." He was spirit; she was nature. As feminists have pointed out, such thinkers projected the ever-changing body they could not accept or control onto women. Of course the male body changes, too; it not only grows and dies, it also includes a penis that often seems to act on its own and has its own body fluids. The unruliness of the penis was blamed on "temptations" of "woman" by most male writers.

In affirming change in the world and in Goddess/God, process philosophy offers a different way to think about the changing body, both the human body and the world body. Inevitably process thinking will also offer new ways to think about women's bodies, men's bodies, and divine body. However, process philosophy did not take up the topic of change in order to provide alternatives to the attitudes toward women and women's bodies in traditional thinking. Process philosophy began as a response to the challenges modern science posed to both the traditional picture of God and to long-standing theological and philosophical conceptions. Feminists, on the other hand, begin with attitudes toward the female body and move from there to the human body, the earth body, and the world body (or the universe as a whole). Though starting from different points, process and feminist thinkers arrive at many of the same conclusions about the changing world and the body. Feminist thinking can make explicit issues concerning women and the body that remained implicit in process thought, while process philosophy can offer to feminism a way of thinking about Goddess/God in relation to evolution. Evolution has not been a priority for feminists in religion. But as thinking individuals in the twenty-first century, feminists also desire a religious philosophy that is compatible with evolution and one that celebrates the wonder of a co-created universe.

Process philosophy is a response to the collapse of the neat picture of creation as eternal and unchanging that had been accepted for centuries in the West.[7] When Darwin proposed the theory of evolution and a growing body of evidence supported it, many people could no longer accept what had been a theological and philosophical consensus for centuries. Darwin alleged that man had evolved from a common ancestor shared with apes, chimps, and monkeys. With this one assertion the whole theological edifice was shattered. Creation did not happen in a single day but in a slow evolutionary process taking millions of years. Creatures did not come into being all at once, corresponding to ideas in the divine mind, but rather had come into being over a long process that appeared to have involved what we would call "trial and error" or "chance." In addition, many species had flourished for a time and then perished. If God had said, "Let there be dinosaurs," for example, this word did not have eternal validity. Humanity was not present at the beginning of the world but appeared rather late on the scene. Humanity was not entirely different from the animals but in fact shared many qualities and capacities with them. We now know that human beings and chimpanzees have 98 percent of the same DNA.

All of this was anathema to traditional thinking. While the death of individuals within a species had been accepted (though the notion of immortality allowed humans to escape even this), the death of species such as the dinosaurs, or the emergence of a new species such as the horse or human beings could not be accounted for. Indeed such a possibility seemed to state that the divine plan for (or eternal ideas of) the world was not perfect in the beginning or to suggest that God didn't know what he was doing. This seemed to be blasphemy—or cause to abandon religion.

Theological opposition to the theory of evolution was vociferous in the nineteenth century and has continued throughout the twentieth century and into the twenty-first, especially in the United States. A few theologians including the Roman Catholics Teilhard de Chardin and Thomas Berry have attempted to transform theology to take account of evolution.[8] Many other modern Jewish and Christian theologians call the creation story of the Bible "myth," yet have not been moved by new scientific knowledge to a fundamental re-thinking of traditional ways of conceiving God's relationship to the world. The language of the Bible and many traditional prayers and liturgies enshrine outdated understandings of creation.

Many thinking people, both men and women, found and continue to find traditional faith to be incompatible with modern science, most especially with the theory of evolution. "How can I go on believing that God created the world in seven days," they say, "when all educated people know that it evolved over millions of years? How can I accept the view that we humans are the crown of creation, when there were millions of years when we did not even exist and it is possible that we might never even have evolved?"

Process thinking takes these questions seriously. Its goal is to transform philosophy in light of modern scientific knowledge, especially evolution, and to provide a new way of thinking about divinity, humanity, and the world (understood as the whole universe) in light of it. The notion that the world as we know it emerged gradually over millions of years of evolution not only challenges the biblical and many other religious accounts of creation. It must also, Whitehead and Hartshorne argued, lead us to question many well-accepted philosophical and theological ideas. The first and most important of these is the negative evaluation of change.

Process philosophy offered another alternative. What if the divine plan for the world was not written in stone at the beginning? What if Goddess/God works in and through the changing world? What if the world was not created, but co-created? This was a brilliant, novel, and radical idea in western thought. If more widely understood, it would allow many more thinking people to accept scientific knowledge and evolution while retaining faith in (a very differently understood) Goddess/God. The idea that Goddess/God acts in and through the changing world requires a thoroughgoing revisioning of the first principles of philosophy of religion and theology. How does the divine power participate in the creative process of evolution? One answer might be that God guides and determines evolution at every step. But this is simply another way of saying that God and God alone created the universe. And it brings us back to the question posed by the extinction of the dinosaurs and other species: Did God make mistakes?

Rejecting this view, Whitehead and Hartshorne propose that the world is "co-created" at every moment by Goddess/God and all the individual beings in the cosmos with a degree of freedom. Process philosophy assumes that the divine power is present within all individuals in the universe, supporting and sustaining them but not determining their every movement. Goddess/God holds the whole together so that it will not explode into sheer chaos. Goddess/God is present with and in every individual, hoping to lure or persuade it to make choices that will be beneficial both to itself and to the whole of life. I will discuss this question more fully in succeeding chapters.

Process philosophy does not imagine a moment of creation when the whole thing got started. Since Goddess/God is understood to be fully relational and fully involved in changing life, it is hard to imagine that there was a time when there was nothing for Goddess/God to relate to. Creation ex nihilo is not accepted. Process philosophy assumes that Goddess/God has always been creative, which means co-creative, and that there has always been something besides Goddess/God in the universe even if only particles swirling in space. This is consistent with at least some current scientific theories about the origins of the universe. Process philosophers say that scientists overstep the boundaries of knowledge (reverting perhaps to theological ideas) if they assert that there was a time, for example before the hypothetical Big Bang, when there was nothing.[9]

It could be said that in affirming the divine presence within the evolutionary cycles of the natural world, process philosophy returns to an earlier vision associated with eastern or indigenous religions or with Goddess religions. Such religions celebrated the processes of birth, death, and renewal as the cycle of life in which the divine power also participates. Process philosophy also celebrates the processes of birth, death, and renewal, and this is not a trivial point. Process philosophy affirms much of the spiritual understanding of the natural world that biblical religions found problematic. But there is a difference. Ancient and indigenous and eastern religions understood life to be a circle, with birth following death and death following birth. Their creators did not have modern science to tell them that life does not only endlessly and wondrously repeat itself, but also, in what to them would have been imperceptible ways, it is continually developing genuinely new possibilities of life. In this larger process, not only individuals but also species come into being and pass away.

Whitehead and Hartshorne state that Goddess/God is the sustainer and ground of all creative processes in the world (and any possible world) as well as a co-participant in them. General principles must structure life on earth, in our solar system, and in the universe as a whole. Yet our knowledge of even the most general principles will always be limited by our positioning in time and space. Whitehead thought (following Plato) that the "forms" that life takes become possible because they pre-exist as eternal ideas in the mind of God. Hartshorne rejected this remnant of Platonism, asserting that the various forms that life takes emerge in the co-creative process of evolution.[10] Hartshorne did, however, speak of laws of nature or limits in which the creative processes in what we know as the universe can exist and continue on. Hartshorne imagined that these include "a basic set of physical laws setting limits to the reign of chance in nature, laws governing the behavior of basic elements, especially hydrogen atoms."[11] Yet even what are thought to be "laws" of nature cannot be perfectly understood. Patterns observable on earth may not hold in precisely the same way on other planets and may not have been exactly the same millions of years ago even on earth. What seem to us to be "laws" of nature might better be called observed patterns or tendencies. Process philosophy takes the careful observation of nature very seriously as a source of information about the basic processes of our world. Still, we must remember that scientific understandings are always changing and developing. What is thought to be a law of nature today may be modified in light of new knowledge tomorrow. When scientists move from description of phenomena and processes to pronouncements about the nature of reality, their opinions should be treated like all other philosophies and evaluated accordingly. We need to be aware that scientists often make philosophical statements about the nature of reality that go beyond the description of observed phenomena.

Process philosophy views modern science as having erred in stating that the so-called natural world is "dead" or "insensate"—that is to say, made up of senseless and unfeeling matter. This view is sometimes called "scientific materialism." In its most extreme forms scientific materialism asserts that all matter—that is, all of nature, including animals and even human beings—act according to scientific laws. In other words, all behavior, human and other, is determined. Freedom and creativity, including human freedom and creativity, are illusory. Some scientific materialists exclude human beings (and perhaps some of the so-called higher animals) from the web of determination. Yet others insist that human beings are no more free than rats in a maze, always reacting according to causes in the same predetermined manner. In recent years this view has been presented by sociobiologists who argue, for example, that human and animal sexual behavior is motivated exclusively by the selfish desire to reproduce individual genetic lineage.[12] Against these views, process philosophy argues that the evidence for scientific determinism is statistical probability, and that it does not prove that there is no freedom or creativity. Given certain stimuli, it can be predicted that 55, 80, or even 95 percent of all rats will act in a certain way. But there is always one rat that will turn left rather than right in the maze, though the apparently known consequence is hunger or an electric shock. Even when dealing with individuals as small as atoms and their particles, scientists have not been able to observe absolute predictability. Process philosophy understands the "uncertainty principle" to leave room for freedom, not only human freedom to react differently to stimuli but the freedom of rats to make choices in a maze and the freedom of atoms to act in unexpected ways. It is this degree of freedom, process philosophy asserts, that creates the grand adventure and risk of life in the universe.

Whitehead and Hartshorne disagreed with scientific materialism's view that the universe is made up of dead matter that acts and reacts according to predetermined laws. Nor was either of them content to consider human freedom an exception to the general rule that everything (else) acts according to predetermined laws. They both felt that scientists were right to desire a unified understanding of all reality in which the same principles could be used to understand the behavior of all individuals in the universe. Against the theory that all matter is "dead," They and Hartshorne made the (from some perspectives) astonishing proposal that all matter is in some sense "alive." Whitehead argued that all individuals down to the smallest particle of an atom have the ability to "feel" and to "feel the feelings of others." In order to avoid the impression that he was attributing humanlike consciousness and feelings to all individuals down to the particles of an atom, Whitehead preferred to use the technical term "prehension" rather than the word "feeling" to describe this activity.[13] As I wish to bring his ideas to a wider audience, I follow Hartshorne in using the word "feeling" to describe the general principle found in all life, while noting that there are obviously widely varying degrees of "feeling" among individuals as different as particles of atoms, atoms, cells, animals, and human beings.

Hartshorne called the view that every individual can to some degree feel and feel the feelings of others and respond creatively to them "panpsychism" or "psychicalism."[14] As the Greek word "psyche" can be translated as "soul," panpsychism literally means that every individual has a soul. Hartshorne said that panpsychism means that there is no part of nature in which "feeling or sentience" could be said to be totally absent[15]; that all individuals can be said to "know or feel or intuit," yet that this must be understood in a "radically broad and nonanthropomorphic sense"[16]; and even that "sympathy, feeling of feeling, the root idea of love, goes to the bottom of things."[17] In other places Hartshorne speaks of "freedom" and "creativity" as inherent in all acts of feeling the feelings of others.[18] Whitehead and Hartshorne both use the word "enjoyment" to describe feeling and feeling the feelings of others.[19] For Hartshorne, feeling, feeling of feeling, sympathy, love, creative freedom, and enjoyment are different yet closely intertwined ways of talking about the fundamental nature of co-creative and relational life on planet earth and, as far as we can know, in the universe as a whole. I will focus on creativity and freedom in this chapter and consider sympathy, love, and enjoyment in subsequent chapters. In attempting to understand what any of these words or concepts means when applied to other than human beings, it should be borne in mind that Whitehead and Hartshorne are not attributing "human" feeling and creativity to nonhuman individuals. Rather they are searching for words to explain their sense that what we know as human feeling and creativity arise out of the web of life and are shared in vastly varying degrees by all other individuals.

In affirming panpsychism, Hartshorne was by no means denying the importance of bodies and embodiment. Hartshorne agreed with many modern scientists that mind and body are an irreducible continuum. He said, "There is much truth in naturalistic materialism. What we would be without bodies is gibberish."[20] Hartshorne further believed that the theory that the soul survives the death of the body is a theological mistake, as I will discuss in chapter five. Thus I feel confident in saying that Hartshorne's perspective might also be called "pansomatism," the view that everything has a body, for the body is the location of feeling. Perhaps the word Hartshorne was seeking is "panpsychosomatism."[21] From this perspective, the discovery that all of our mental processes have bodily correlates is not proof—as is sometimes asserted—that we are nothing but bodies, nor is it proof that there is no feeling or creative freedom. Rather, such discoveries show that we think not with our heads or brains alone, but through our bodies.

We would not feel any sensations in our minds unless our bodies continually sent messages to the brain. Humanistic psychology—for example, gestalt and bioenergetics—confirms that the mind and body are one system. Thus therapy requires not only talking, but also unlocking the feelings stored in the body. Healing bodywork takes this one step further into a focus on the body as the locus of blocked feelings. However, healing bodywork will usually incorporate talking, so that connections can be made with habitual ways of thinking. The notion of a mind-body continuum finds further support in recent discoveries of biopharmaceutical treatments for so-called mental illnesses that were once considered to have solely psychological causes—that is, to reside solely in the mind. Holistic medical practice, on the other hand, is increasingly recognizing the complex ways in which mental attitude affects illnesses once thought to have solely physical causes.

For process philosophy, the ability to feel and to feel the feelings of others to some degree, or to relate to others in some sense with creative freedom, with sympathy, with love, and with enjoyment is to be attributed to all individuals in the cosmos. This raises the question: What does process philosophy mean by an individual? Unitary or singular individuals are those who can feel. These include human beings, animals, cells, atoms, and particles of atoms. Plants, rocks, and mountains are not individuals but societies or aggregates of individuals, according to process philosophy. Most of us have little difficulty understanding that animals are feeling individuals to one degree or another. Hartshorne asks us to consider that even the cells of our bodies have feelings, which they can communicate to our brain as pleasure and pain. The cells of our body operate to a greater and lesser extent apart from our conscious will in the processes of sleeping and waking, digestion, healing, and illness. How can this be possible if our cells do not have a capacity to respond creatively to each other? The more I read about the process of evolution, the more likely it seems to me that cells do have some level of feeling and creativity.[22] True, it may be hard for us to imagine what kind of feeling a cell, much less an atom or a particle, might have. Still, statistical uncertainty suggests that even atoms and particles do have a small degree of freedom, and therefore the possibility is left open that they also have feeling, although to a very limited degree. The process view does not contradict the empirical findings of modern science on this point.

So far I have discussed feeling in animals, cells, atoms, and particles, but not plants or rocks or mountains. Do these too have souls or feeling? The answer given in process philosophy is that plants are societies of individual cells working together. The cells in a plant have feeling, while the plant as a whole does not. Process philosophers state that plants do not have a single organizing center or brain in the same way that cells and animals do. For example, plants can lose limbs without apparent loss of vitality, many plants regenerate themselves from roots alone, and whole plants can be recreated from cuttings. Plants do not have the same relation to their "body parts" as animals. We can survive the loss of an arm or a leg, but we cannot regenerate ourselves from our feet or even from our brain (though some scientists believe they will soon be able to clone people from cells). Hartshorne reiterated his view that plants are a society of individual cells in the letter I received from him shortly before he died.[23] According to the process understanding, minerals are individuals on the smallest level. Mountains and rocks are not individuals nor even societies, but aggregates of individual molecules for a greater and lesser time joined together.

For Hartshorne the understanding that plants, rocks, and mountains are not individuals meant that "it is idle to try to empathize with a mountain, and probably even with a tree."[24] Indeed it was my writing to him that I was singing to the plants in my garden that provoked Hartshorne to respond with his view that plants are not individuals. Martin Buber, in contrast, wrote that he had an I-thou relationship with a tree and a piece of mica. My experience is more like Buber's. Yet Buber remained agnostic on the question of what this experience meant. "Does the tree then have a consciousness similar to our own?" he asked. "I have no experience of that,"[25] he replied to his own question. I am not certain that the process view of plants as societies of individuals is sufficient to explain the degree of cooperation among the cells of the plant. However, whether one considers plants as individuals or as societies of individuals, the consequences for life are the same. For whether it is the cells of the plant or the plant as a whole that is a living being, the vegetable world is alive. And thus loving our plants or singing to them may indeed make them flourish,[26] even if the response is on the level of individual cells rather than on that of the plant as a whole. Similarly the individuals that make up a rock or a mountain are alive. And perhaps they too can be communicated with. For me this is the logical conclusion of the process view that there is nothing in the universe that is "mere matter," existing entirely to be used.

For process philosophy, the notion of panpsychism helps to explain the evolutionary process. If matter is "dead," then it is hard to explain the transition to "life." And if only humans or only humans and some animals have the ability to feel, then it is difficult to imagine how humanity evolved out of the web of life. The Roman Catholic evolutionary theologian, Teilhard de Chardin, found this a problem as well. He asserted that there were jumps or gaps in evolution that could not be explained. Not only the transition from non-conscious to conscious life but also the transition from so-called inert matter to living organism could not be explained, he thought, by science alone. Teilhard hypothesized that at these points of transition, the divine power added an extra surge of creativity that enabled what he saw as one kind of thing to evolve into something quite different. In this way, creation and evolution could be harmonized.

Process philosophy argues that it makes more sense on rational as well as experiential grounds to imagine that all individuals in the universe are in some sense living, in some sense able to feel, in some sense free and creative. On this view, no one has to explain how living beings evolved from non-living beings or how creative and feeling individuals evolved from uncreative and unfeeling ones. From a process point of view, the unexplainable gaps between living and dead matter and between intelligent and unintelligent life proposed in classical evolutionary theory do not exist. Instead relatively more creative and feeling individuals evolved slowly and through many stages of development from those with relatively much less creativity and feeling. This process occurred over millions of years in gradual stages. Process philosophers argue that their view better explains the process of evolution than other theories.

While some modern scientists and many traditional western religious thinkers have asserted that human beings are the only animals with consciousness and feelings, this view is contrary to the experience of children, poets, artists, mystics, and many other adults. Hartshorne's notion of panpsychism provides a way of thinking about a wide variety of experiences that otherwise might be termed inexplicable or irrational. Most people who have pets know intuitively that they have feelings and intelligence. This is increasingly being confirmed by intensive studies of animal behavior.[27] In the novel *Chocolat* an old man suffers deeply with the suffering of his aging dog, and the priest's declaration that he should not worry so much about an animal who has no soul only intensifies the man's pain.[28] A theology that understood there to be a continuum of feelings and intelligence between human beings and animals, and one that asserted that the divine power cares about all creatures, would have been a blessing to the old man and his dog. Like many others, I have always felt that not only animals but the whole world is alive. I held on to this feeling when I grew up, though I had to learn to defend it against more traditional thinking that called it childish. Some retain their childhood outlooks in secret, not telling anyone so as not to be made fun of. The view that the world is alive is common among indigenous and some non-western peoples. Process philosophy provides a way of thinking about these experiences.

Process philosophy understands that all change in the universe is the result of the exercise of creative freedom. Freedom is not just the ability to turn left rather than right. Nor is it sheer freedom, the ability to do absolutely anything. Freedom exists within limits that include the patterns and possibilities found within nature and even more important, the other individuals in the world who are related to us. At the most fundamental level freedom is the ability to make a new creative synthesis out of elements that previously existed, but not in the same combination. The adage "each day is a new creation" is compatible with process thinking. Process philosophy states that each instant is a new creation for every individual in the universe.

As I sit here writing, I am continually ordering my thoughts; in fact, I had not thought of the sentence I am writing now until I began to write it. This sentence is a new creation, not copied from an eternal store of sentences in my brain or in the brain of Goddess/God. Is it an entirely new thought? No. Other writers have probably written similar things, and the train of my thought in this paragraph owes a great deal to Hartshorne and to my own previous writing. I would probably not be writing this sentence at all (though I might be writing another one) if my parents had not sent me to college, if I had not taken a certain philosophy course, if I did not have a desk and a computer waiting for me, and so on. In this sense my sentence is also co-creation. My sentence is not entirely novel, nor is it mine and mine alone. Still, in this moment or more precisely, in a moment that is already past, it is my sentence, a creative synthesis much like the one you are making now as you read this sentence and try to figure out how it applies to your life.

We tend to think of creativity as something some people have and others don't. Creativity is for artists and writers and filmmakers. Most of us just aren't that talented, we are taught. We tend to think of artists as disembodied spirits, who, like the traditional God, create something out of nothing. In fact creation is always out of something. Creation is not a capacity within individuals but rather a relationship between an individual and the world. In this sense creation is always co-creation. Some potters speak of form emerging out of interaction with clay. Michelangelo discovered forms emerging as he worked with specific pieces of marble. Recognizing creation as co-creation makes it less mysterious. As a child, I was told that I had a special artistic talent. Yet I discovered that even the "hoods" and "toughs" who were assigned to art class because they had difficulty with reading and writing had artistic talent. I was told not to bother with singing, because I "couldn't carry a tune." Imagine my surprise—some forty years later—when my Greek friend with the beautiful singing voice told me my voice wasn't so bad and that I could learn to sing even better. After a lifetime of avoiding singing, I now find that I can sing, perhaps not as clearly and sweetly as my friend, but with enthusiasm and grace that are sometimes appreciated by others.

Process philosophy tells us that co-creative freedom is something that all individuals exercise every moment of our lives. The mother who makes up the words or the tune as she sings a lullaby to her child is creating a new song. The child who plants seeds in a garden is creating beauty where perhaps there was none before. The grandmother who decides that this time she won't spank her grandchild but will try to reason with her is creating a new way to interact with children. The rat in the maze is making a new creative synthesis when it figures out that going right means food while going left means electric shock. After that, it makes a determination each time it is put into the maze, as to where it will go. If process philosophy were more widely understood, there would be a lot fewer rats in mazes. A rat in a field feeding its young has far more choices and a far more interesting life than a rat in a maze. We are able to torture animals in part because we have been taught by both traditional religion and classical science that animals do not have consciousness and feelings analogous to our own. Process philosophy also suggests that to a lesser degree, creative synthesis is occurring in the cells of our bodies and even in the smallest particles of matter as they interact with other particles.

The exercise of creative freedom does not always have positive results. Hitler made a new creative synthesis when he activated long-standing currents of anti-Semitism and combined them with Germans' feelings of inferiority after having been defeated in World War I to create a new program of violence against Jews. His use of modern technology in his program of hatred was also a new creative synthesis, one that changed the face of the world for the worse. Freedom can indeed be used to create enhance life and to destroy it. I will consider this more fully in chapter four.

Process philosophy asserts that creative freedom is one of the fundamental principles of the universe. It is therefore appropriate to understand divine power as co-creative freedom. The creative freedom of Goddess/God in process thought is not that of a God who exists alone and created the universe out of nothing. Rather it is the co-creative freedom of a divine power involved with all other individuals in the universe. In order to assert the intimate relation of the divine power with the world, Hartshorne makes the startling assertion that the world—meaning the cosmos as a whole—is the body of Goddess/God. Hartshorne claims that this idea was anticipated in Plato's late work the *Timaeus,* in which Plato suggests that the world-body has a soul.[29] However that may be, Plato's earlier and more consistently expressed idea that the Good must be separate from change and the body is the one that was built upon for centuries in the west. The notion that the world is the body of Goddess/God is related to ancient and contemporary understandings of Earth or Mother Earth as the most ancient of the Goddesses. Similar ideas may be found in some non-western traditions. The modern "Gaia hypothesis" views the earth as a unified system.[30] However, the process view can be distinguished from these views in its insistence that it is not just the earth but the universe as a whole, also called the world by process philosophers, that is the body of Goddess/God. The earth is holy because it is part of the body of Goddess/God, but the body of Goddess/God is vaster than our earth.

What does it mean to say that the world is the body of Goddess/God? At bare minimum it means that everything that happens in the world happens to Goddess/God as well. Everything we do or suffer, everything animals and cells and molecules and atoms do or suffer, whether on this earth, on Mars, or on a distant star, becomes part of the experience of Goddess/God. Our experiences are as intimate to Goddess/God as the experiences of our bodies are to us. Since the body of Goddess/God is the changing world, Goddess/God is fully involved with the changing world. Hartshorne understands the world as the body of Goddess/God on the analogy of the relation of our bodies and our selves. He is convinced that there can be no psyche, no soul or feeling center without a body. As he says, "The human soul as disembodied is hopelessly unclear and false. A merely disembodied God is an unfounded idea."[31] As stated earlier, process philosophy agrees with modern science that soul or mind and body are one system. Hartshorne takes this understanding one step further in applying it to God. The principle of embodiment is true of Goddess/God and not only ourselves. Goddess/God cannot be an exclusively spiritual or mental being. Goddess/God could not exist without a body. Goddess/God is the soul or consciousness and organizing principle of the world body. As the world body changes in large or small ways, so Goddess/God must also experience change.

This certainly makes it clear that Goddess/God is not separate from the world. But doesn't the conception of Goddess/God as the soul of the world body take away the freedom of the rest of the individuals in the world that process philosophy has worked so hard to assert? In answer to this question, Hartshorne asks us to consider our bodies as a society made up of cells that to some extent operate independently of our conscious will. People don't tell their stomachs to digest the food they have eaten. We don't tell a finger to bleed when it is cut, and most of us couldn't even begin to tell a finger what it needs to do to heal. We don't tell our blood cells to fight off the free radicals that can cause disease. These things happen on the cellular level and are fully or largely unconscious. Let us say that these things occur because of the will of the cells.

In other respects the cells of the body interact with our consciousness. When we cut a finger, our cells send a message to the brain that gets recorded as pain. This causes us to attend to the wound. When the sun is shining, the cells of the body may send a message to the brain that is recorded as joy. This may cause us to go outside and take a walk. The conscious mind also interacts with the body. Right now I am telling my fingers to tap on certain keys of the computer. A moment later, I pause to look at the bird called a great tit that is hopping around on my balcony at the edge of my view. Then I direct my attention back to the computer screen and my fingers move again in accord with the words that form in my mind. Before too long the cells of my eyes or my back will start telling me to get up and move around, or the cells of my stomach will tell me to eat. Psychosomatic illnesses can be understood as the result of communications from the mind received and acted upon by the cells. Some alternative healing methods are ways of training the mind to give health-inducing messages to the cells of the body.

The relationship of Goddess/God to Goddess/God's body is something like our relationships to our bodies. There are times when the members of Goddess/God's body act relatively independently of the will of Goddess/God. However, they still are part of the whole, and in this sense are part of the body of Goddess/God. There are times when the members of Goddess/God's body communicate directly with Goddess/God or when Goddess/God communicates directly with one of them. This, however, is an analogy. We cannot cut open the world body and find the brain of Goddess/God. Nor do we imagine that Goddess/God was born and will die.

The process view of the world as the body of God has as much in common with pantheism as it does with traditional theism. Traditional piety places God "above" the world. Classical theists influenced by Plato's *Symposium* imagined God as existing entirely apart from the changing world. The problem with these views, as we have seen, is that they lead to too great a separation of Goddess/God from the world. Pantheists, in contrast, understand the world as divine. Pantheists say "all is God" or "God is all." As a mystical perspective on experience, pantheism leads us to appreciate everything in the world because everything is understood to be sacred or divine. Process philosophers agree that everything is sacred insofar as everything is in the body of Goddess/God. However, if "all is God" or "God is all" are taken as philosophical statements, the implication is that divine power is the only power in the universe and that any power we have is without anything left over simply a manifestation of it. In other words, I am not typing this page, God is. I cannot be given credit for the good that my words may do nor be asked to take responsibility for any evil they may create, because in the end "I" do not really exist; only God does. Process philosophy asserts that pantheism leaves no room for genuine individuals and no room for freedom.

In contrast to both theism and pantheism, process philosophy offers "pan-en-theism," the view that "all is in Goddess/God."[32] More specifically for Hartshorne, everything is "in" the body of Goddess/God which is the world or universe as a whole. Panentheism affirms Goddess/God's intimate relationship to the world while affirming individuality and freedom. For process philosophy, individuals are in the body of Goddess/God, but we also can enter into relationship with Goddess/God because we also exist as individuals. Again, think of the analogy of cells in our bodies as individuals and as part of a larger whole. Panentheism affirms the sense of unity or connection affirmed by pantheists while retaining the sense of differentiation and individuality affirmed by theists.

To say that the world is the body of Goddess/God is to say very clearly that Goddess/God is not separate from the changing world. Does this mean, as Plato feared, that Goddess/God can become more or less? And if so, could Goddess/God cease to exist? To answer this question, Hartshorne introduces what he calls the principle of "dual transcendence."[33] Goddess/God is said to be changeable in one way and unchangeable in another. Another way to put this is to say that Goddess/God's nature or character does not change, but that Goddess/God's experience does. It is the unchangeable nature (also called the primordial, abstract, or absolute nature) of Goddess/God always to exist and always to act and react in a consistent way—in other words, always to be creatively and lovingly related to a world. The reasons for considering the divine nature to be loving will be discussed in the next chapter. In being unchangeable in basic nature and character, Goddess/God is different from all other beings in the universe. To the extent that finite beings are free, we may choose not to act in consistently loving and creative ways. We also will surely die. Goddess/God does not change in these ways. Therefore, Goddess/God has an unchangeable nature as traditional thinkers asserted. But this is only part of the story. Goddess/God is involved with a changing world, and therefore Goddess/God can and does change. What can change in and for Goddess/God is experience (also called Goddess/God's consequent, concrete, or contingent nature).[34] Goddess/God's experience changes, depending upon the experiences of the creatures to which Goddess/God is related. Dual transcendence is such a simple and logical idea that it is astonishing that western philosophy and theology struggled so long without it.

The process view that Goddess/God really is involved in a changing world, means that in some sense Goddess/God can become more or less than Goddess/God was before. As long as it is understood that it is Goddess/God's experience that is changing and not what we have called Goddess/God's character, this is not a problem for process philosophy. Process philosophy asserts that Goddess/God truly does rejoice in our joy and truly does suffer with our suffering. When individuals in the world are suffering, Goddess/God is suffering too. When suffering increases in the world, the suffering of Goddess/God increases. There is no question that Goddess/God would prefer to experience joy rather than pain and non-productive or harmful conflict, and, in that sense, Goddess/God's experiences do become "less" when suffering increases. This is the risk of creative freedom. This risk is greater than once might have been thought. Some decades ago many people thought that although individuals certainly suffer, it made sense to think that the earth as a whole is evolving in a positive direction. Indeed many people—including scientists and politicians, as well as New Age thinkers—still hold this view. However, in light of the widespread suffering and environmental degradation characteristic of our times, it seems entirely possible that the planet as a whole is not getting better but rather is getting worse, not only for human beings, but for many other species, some of them already extinct. The air, the sea, and the dry land are being poisoned. The cause of this is the choices human beings are making individually and collectively with our creative freedom. If things really are getting worse on the earth as a whole, the suffering of Goddess/God must be far greater than we can imagine. The significance of this greater degree of risk for the future of Goddess/God, for the human species, and for the world will be considered in chapters four and seven.

Process philosophy offers the West a radically new way of thinking about divine power and its relation to the changing world and with it a new way of viewing human beings as in the image of the divine power. For process philosophy, "change is." The first theological mistake—God as perfect and therefore unchangeable—is overcome. Change, freedom, and the mind-body continuum are understood together. Without bodies there would be no change. Without change there could be no freedom. Creative freedom is how embodied minds respond to the changing world. We are our body-minds, and the same is true of Goddess/God, whose body is the world or the universe as a whole. Goddess/God is involved in the changing world and responds to it with creative freedom. Is it possible that this was not understood in so many earlier philosophies and theologies because the changing world and the body were identified with women whose secondary status was unquestioned?

Chapter Three

Touch Is

In the process view, the world is a web of changing individuals interacting with and affecting each other, co-creating the world. Relationships are the building blocks of life. In them we grow and develop. In them we experience the joy of living. In them we are traumatized and suffer. Without them, we would not be. Personal relationships are embedded in a web of structural relationships that shape societies and cultures. The world is social through and through. When others are suffering, we suffer. When others are happy, we feel their joy. We are closely related to individuals in our own species, more closely to some than others, as are all animals. Many of us are related to animals we care for and to the plants (or cells of the plants) in our fields or gardens. Consciously or not, we are intimately related to the cells that make up the food we eat. Consciously or not we are related to the constellations of minerals and cells and animals that make up the environment in which we live. Landscapes and places as well as people color and shape our experiences of the world. If the air we breathe is clean, we may breathe more deeply. If it is poisoned, we may gasp for breath, cough, and eventually become ill. Atoms and cells also exist in relationships. Atoms touch each other and become rocks or mountains. Cells touch each other and become plants. The web of life is no mere metaphor. We are not only embodied, but embedded in a web of relationships. For process philosophy, Goddess/God is also embodied and embedded in a web of relationships. Goddess/God touches and is touched by every individual in the universe. This is what is meant by Goddess/God's "sympathetic goodness." Goddess/God is "supremely related to all."[1]

Relationship, relationships. Few of us can hear those words without an added emotional charge. "When you get older," we were told, "you will fall in love and get married." As teenagers and adults, we all want to be in love, to have "a relationship." When we find that special someone (whether we are straight, gay, or other), we feel that we have fulfilled our dreams and the dreams others had for us. If we don't have that special love, we feel badly about ourselves. "I don't have a relationship right now," I have said tearfully, more than once, ignoring the relationship I had at that very moment with the person listening to my words. Not to mention the relationships I had at that very moment with the glass of red wine in my hand, the blue velvet chair on which I was sitting, and the music of Judy Collins that was playing in the background. Not to mention the parents who brought me up and still cared about me, the many friends I had, and the students and colleagues with whom I interacted as a university professor. Appreciating all our relationships puts things into perspective.

There is another problem with the way we think about relationships. "Relationships are a girl-thing," boys are told. "Women and girls are good at them. It is their sphere. Leave it to them. Men and boys have other and more important interests: sports, politics, making a living. Women cry when they are hurt. Men don't. Caring too much about relationships is a weakness. A man must be strong and independent." This is a story we have all been told over and over again in more ways than we can number. Relationship and independence are contrasted with each other. One is for women, the other for men. Women who want to be free often feel (or fear) that we must learn "to act like men." Men who put relationships first wonder if others will think they have "compromised their masculinity." Many career women put sexual relationships and family second or on hold in their lives, while others struggle to have it all—often with male partners who themselves have learned to put relationships second. Habits of mind are very hard to break.

The notion that relationship is a weakness can be traced back to Plato and the traditional thinking that followed his lead. Besides saying that the Good cannot change, Plato also said that the Good is "absolute, existing alone with itself, unique, eternal."[2] The idea of the Good or God as "existing alone" is contrary to the biblical picture of God involved with the world, creating, loving, and judging it. So why did Plato and the philosophers and theologians who followed him assert that God exists alone? Because they viewed the risk involved in relationship as a potential liability. If God cares about others, God will be happy when they are happy, sad when they are sad. This means God's happiness can be increased or diminished. It would not do for God to become more or less, because God is already perfect. If God is to be absolute or reliable, he must be absolutely independent, absolutely unrelated to any other being in the universe. For this God there can be no touch. This is the traditional philosophical and theological view in the West, strange as it may seem. Hartshorne called it the common theological mistake of asserting God's unsympathetic goodness. "No one could really believe that God is unsympathetic," it might be responded. "This emperor really does not wear clothes!" That may be, but this is the God I met when I studied the classical tradition in western theology and philosophy. I often wondered how so many great minds could have gotten it so wrong.

The unrelated, untouchable God became a model for humanity as well. For centuries the Roman Catholic tradition has dictated that priests must not marry and Christian Orthodoxy has allowed only celibate men to rise in the hierarchy of the church. For centuries women and men in eastern Orthodox and western Christianity chose the monastic life, believing that freedom from intimate relationships would allow greater room for God. Some Jews also have been attracted by asceticism, though no monastic traditions developed. This attitude is not limited to the West. Both Hinduism and Buddhism have long-standing ascetic traditions.

In the modern world, celibacy ceased to be the primary issue. Nor was God any longer in the foreground. Yet modern thinkers still touted human freedom as independence. Independence of what? Of relationships with others and the physical world, including the body. When Descartes said, "I think, therefore, I am," he was not breaking with Platonic tradition but reasserting it for the modern world. The thinking individual imagined by Descartes knew only one thing for certain: that he thought. He did not know with equal certainty that he had a mother, a father, or brothers and sisters. He did not know with equal certainty that he had a body that was sitting in a chair or an arm that rested on a desk as he wrote, or that he wrote his words with a quill pen. For Descartes these things all had a degree of uncertainty. The self (as he understood it) was defined by his own mind; the self (as he understood it) was independent, existing alone. Relationships were not understood to be a given in life but rather something to be chosen or refused. The obvious fact that human beings live in communities and societies that precede and survive them was a problem for thinkers who followed Descartes. To solve this problem they came up with the idea of a "social contract."[3] Human beings were said to have "freely entered into" society with each other, presumably not having been related to each other in any but the most barbarous ways previously. It was because philosophers were still having trouble getting the thinking self of Descartes into relationship with the world that a paper titled "A New Proof for the Existence of Other Minds" was not considered out of the ordinary in my graduate program.

These attitudes have filtered down into ordinary life. Most of us unconsciously think of writers, artists, and composers struggling alone in garret rooms—even though many of them have comfortable lives as well as wives or husbands and children. We are used to thinking that having the freedom to create is incompatible with personal relationships and physical comfort. In the same way many people automatically think that women's freedom requires leaving home and family behind. This was a theme of a number of early feminist novels and autobiographies, and it is still the way media often portray feminists. Some women are afraid to call themselves feminists because they assume women cannot be both free and related to others.

In the United States, and to a lesser extent in other democracies, freedom and independence have been exalted above all other values. Traditional liberals argue that free speech is an absolute right. The individual has the right to speak his or her mind, no matter what the consequences to others. Those who counter that some speech creates a social climate in which others (women, African Americans, homosexuals, Jews) are inevitably harmed have a more social conception of reality, in which individual freedoms must always be weighed against the rights of others. These two groups cannot hear each other because they begin with very different understandings of the self. One takes the freedom and independence as primary, while the other sees every self as embedded in a web of relationships. On the other side of the political spectrum, conservatives argue that the right of free enterprise is absolute. No one should stop anyone from making money any way he or she wants, and no one should take that money from him or her in the form of taxes to be used for those who cannot make their own way. Those who see the need for social welfare programs to aid the less fortunate begin from a more social conception of reality. In it, the individual freedom of some has been enhanced by social positioning, while that of others has been limited, and social programs can make up at least part of the difference. Whether social programs could be designed better is another question. Notions of the individual independent self untouched by relationships, as well as more social views, can be found on both sides of the American political divide.

But what if relationship and freedom are connected? Like two sides of the same coin? Process philosophy offers a social view of reality in which all individuals in the world, human beings among them, are connected to others, to their immediate environments, to social worlds, and in a larger sense to the whole world and all beings in the web of life. For process philosophy, creative freedom is co-creative, always in relation to someone or something. Individuals with creative freedom are embodied and involved or engaged in a web of relationships. Creative freedom makes new relationships and makes existing relationships new.

For process philosophy, as for Jewish theologian Martin Buber, there can be no "I" without a "thou," no self apart from relationship. Martin Buber said that before speech is developed, the hand of the infant reaches out for its mother (or other nurturer).[4] In other words, before Descartes could formulate a thought, and certainly before he knew that he thought, he reached out his hand in relationship. The existence of the other is as certain as the existence of the self. Long before infants learn to speak, they come into relationship with others besides the mother, and with the physical world, with cribs, toys, sunbeams, shadows of leaves blowing in the wind. The existence of a world and the existence of others can be doubted only by someone who imagines that he or she could exist apart from relationships. According to process philosophy, a person who imagines he has no relationships is to be pitied—or committed to a mental institution. His thoughts on this matter certainly should not have become the foundation of modern western thought.

For process thinking, relationships are not antithetical to the development of the self; they are necessary to it. Good parents are those who encourage a child to grow and develop and exercise its own freedom of judgment, even when its ideas or actions are not exactly what the parents might want. Good friends are like this too, as are good lovers, and partners of all kinds. Freedom is not antithetical to good relationships; it is fostered by them. This is not to say that there are no limits. Good parents do not encourage Johnny to exercise his freedom by beating up on his little brother. They would not let him ride his bike on the freeway. But they would allow him decide to become a dancer or a plumber rather than a doctor, if this is what he wants. And they would accept him whether he chose to marry, enter into a gay relationship, or remain single. This is not to say that there are never conflicts between self and other. Of course there are. We all have interests that sometimes compete or conflict with the interests of others. Nor is it true to say that we never experience conflicts between our desires to be free and independent and our desires for certain relationships. Most of us know this all too well. Sometimes we do have to end particular relationships that are damaging to ourselves or others.

Still, it is a mistake in thinking, encouraged by our culture, to imagine that we can free ourselves from all ties. When we end relationships we turn to friends and family for support. Sometimes we try to cut ourselves off from all others or feel that we have been cut off from them. But in fact we are still embedded in many webs of connection, starting with the air we breathe. In the choreopoem *for colored girls who have considered suicide when the rainbow is enuf,* a black woman looks up from her despair and sees a rainbow. For her the rainbow really is enough. She had lost everything she felt was important, including her children, but the rainbow reminded her that she was still alive and linked to great beauty outside herself.[5] The reason we often feel that we would rather die than be cut off from all relationships with other human beings is that we know deep down that they are the stuff of life. Our culture has done us a great disservice by causing us to focus on one relationship to a life partner, often to the exclusion of all others. I certainly made a mistake in thinking when I said, "I don't have a relationship right now." That mistake caused me a great deal of suffering. It made me feel alone in the universe, when in fact I was not.

Process philosophy says that relationships are "internal," not "external." This is a technical philosophical way of saying that relationships really do matter, really are the foundation of all that we are. Some have imagined that the self is solid, like a rock or a fortress. "Don't let anything bother you," we say. "Don't let them get to you." The self may be connected to other things and other people, but at least when we are strong we imagine that there is a core of the self that is impermeable. We say that we "lose" this self when we fall in love or when we care too much about anything, whether it be a progressing in a career or winning a game. We say we lose control for a time, but then—if we are lucky—we "come back" to this self. When philosophers think this way, they say that there is a substantial (that is, rock-hard) self that stays the same no matter what happens. Relationships are external to this core of self. They come and go, while it stays the same. The self has been understood as hard and substantial by most philosophers in the West.[6] The individuality and freedom of the hard and substantial self is assured, but it is isolated from others. Others never really touch its core. How such a self can enter into relationships becomes a "problem" for philosophers.[7]

For process philosophy, relationships are a given. The self is created in relationships and it is deeply affected and changed by them. Relationships are internal; they affect and change us at the most basic level. After I fall in love, I will never again be "the same person." I will have been changed by what I have experienced with my lover. Even if I didn't, as we often say, "learn from my experience," and thus continue to fall in and out of love in many of the same ways with others, I am not simply repeating a pattern—though I may be doing that as well. Every experience is different, and at the very least, after the second relationship, I am someone who has been in love more than once. The ancient Greek philosopher of change, Heraclitus, said, "it is impossible to step into the same river twice." This is not only because the river will have changed and the person stepping into it will have changed, but also because both the person and the river will have been changed by each other.

When I saw six black storks in the local reservoir at the end of a recent summer, I was astonished. Not only because storks wearing black and white tuxedo feathers teetering on bright red legs while dipping long red beaks into water are beautiful in a way that never quite becomes comical, but also because I had never seen them there before. When I went back the next day and they were still in the reservoir, I was pleased but not surprised. I was not the same because I had experienced their particular beauty and because I expected them to be there. Presumably they were not the same either, because after stopping near my Greek village on their annual fall migration (possibly for the first time), they had discovered that they liked it there enough to stay for a few days.

One could say that these changes are trivial. I was still a birdwatcher and they were still storks and the reservoir was still full of water. Eastern thinkers and process philosophy encourage us to think differently. The reservoir had less water in it the second day than it did on the first in part because storks had been wading in it. This change may have been imperceptible to me, but it was nonetheless real. The storks were eating something that was in the reservoir—perhaps frogs or small fish. If so, there were fewer frogs or small fish in the reservoir than on the previous day. The storks may have looked the same to me, but for all I know, they were coding in their brains the information that this reservoir is a good place to stop, something they didn't know before. Next year they might anticipate it, whereas last year they may not have noticed it while flying over.[8] And me—surely there is a core of me that is the same. I am still a birdwatcher. But I wasn't a birdwatcher three years ago. I have always loved nature—in that way I am the same. And yet, birdwatching has changed me. I know that if I go outside and find the places where birds are, I can always be happy. This is a new feeling. I now teach my friends how to watch birds and seek friends who enjoy birdwatching too. In years to come, I will be back at the local reservoir with them, looking for black storks. It could be said that it is all a matter of perspective. In some ways the reservoir, the six black storks, and I are the same, in others different. But perspective leads to two different ways of looking at the world. One affirms change and asserts that we are deeply and really changed by our relationships. The other says that we do not change in essential ways and that nothing really affects the core of self that is ever the same. External relations are those that do not significantly change us. Internal relations are those that do.

Whitehead and Hartshorne were very much aware that they were departing from much of western thinking and from views widely accepted by philosophers who were their contemporaries when they argued that internal relations change the self. In order to explain how relationships change individuals, Whitehead broke individuals down into momentary units of experience. He then stated that units of experience, each different from the next, are more fundamental than individuals. To understand what he meant by this, recall Walt Disney's movie *The Living Desert.* Remember how the camera was set to click at intervals while the desert flowers were opening. What the audience saw was a single flower opening. The camera on the other hand captured separate images. Were there many different flowers or only one flower? Ordinary understanding says there was only one flower. But what if the bud had been blown off in the wind? It would never have become a flower. A flower must once have been a bud, but a bud will not necessarily become a flower. Whitehead and Hartshorne say that our experience is like the images captured by the camera. Each instant we, so to speak, create ourselves anew. Of course there is also a sense of continuity. The new flower or person has a memory of the one that just preceded it, for they inhabit the same body. Whitehead and Hartshorne were well aware that the view of the self as a series of fleeting moments can also be found in Buddhism. Both Buddhism and process philosophy anticipated the deconstructionist criticism of the substantial or essential self.

I agree with criticisms of the substantial, unchanging self. I have long known that relations are internal, affecting us deeply, causing us to become different. I sense that we are changing all the time as we and our relationships shift. But I am not certain that it is necessary to break experience down into discrete momentary units in order to assert this. When we look back at past experiences—our own or someone else's—we can see breaks or gaps. In many ways I am not the girl in an organdy dress who climbed my grandmother's peach tree, nor the young woman who felt mystically transported in Notre Dame cathedral in Paris, nor the not much older one who studied theology at Yale. Yet I do not experience my life as being made up of instantaneous fragments of experience that I hold together by somehow creating a new self in every moment. For me experience is more flowing, more like "a continuum [in which] no definite single parts can be found."[9]

Hartshorne admitted that life feels more like a flow than discrete fragments,[10] yet he maintained that experience really occurs in discrete fragments. The question is why Hartshorne and Whitehead adopted a position that seems to many to be counterintuitive. I suspect that Whitehead found it necessary to break experience down into discrete momentary fragments in order to explain how relationships can change the self for two reasons. One was that he wanted to make his philosophy of the self conform to a mathematical model. The notion that the self could be reduced to small momentary independent units of experience probably appealed to his mathematically trained mind. The other reason was that both he and Hartshorne were arguing against a philosophical tradition that had consistently imagined a self so substantial that its relations with others could not be said to affect it. Both Whitehead and Hartshorne were influenced by and responded to the work of the mathematician, logician, physicist, and metaphysician Leibniz, whose theory of "monads" influenced many other philosophers. Leibniz defined individuals (human and nonhuman) as monads so tightly constructed that they could not be influenced in any significant way by their relationships with others. Their relationships therefore are external, not internal.[11] When I studied Leibniz, the monads were compared to billiard balls that might collide but whose hard exteriors could not be penetrated. This was not a theory of the self that ever appealed to me. Nor did this aspect of Leibniz's thinking appeal to Whitehead and Hartshorne. Yet because Leibniz's theory had been so influential in the history of philosophy (and because Whitehead and Hartshorne saw themselves as building upon the work of other philosophers), they took Leibniz's theory as the starting point from which they would begin to imagine an alternative theory of the individual.

If, as I suggest, Whitehead and Hartshorne were responding to an understanding of the individual unrelated to others and with boundaries as hard as a billiard ball, then only by "breaking" this billiard ball into fragments could they assert that it could be changed by relationships. As I see it, the notion of the self as made up of discrete moments of experience was a response to problem that itself was created in the history of philosophy.

Hartshorne also found the notion of experience occurring in discrete fragments necessary to explain altruism. Only if my future self is no closer to me than the future self of someone else, he reasoned, would it be possible to understand how we could "love our neighbor as ourselves."[12] I believe that when they constructed their theories of the self, Whitehead and Hartshorne gave too much weight to the history of philosophy (and in Whitehead's case also to mathematical precision) and not enough to life. I do not think we need to take Leibniz as a starting point, nor mathematics as a model. Reflection on ordinary experiences such as interest in the arrival of black storks at a reservoir should be sufficient to make the case that the self is changed by relationships. Reflection on experiences like friendship or parenting should be enough to convince us that it is possible to love another as oneself. Thus I believe that we can understand relationships as internal and the self as continually changing without accepting the Whiteheadian view that the experience occurs in discrete fragments. However, Buddhists will find the Whiteheadian view compatible with their own.

Does the view that the self is always changing mean there is no core of the self, no identity? Buddhists say that when we do not recognize and expect this, we "cling" or "grasp." We cling to visions of ourselves or others: "I would never do that," or "She is always like that," we say. Or we hold on to a notion of the way things are. "We are a happy family," we say. "There are no problems here." Or to a vision of how we want things to be. "We deserve happiness," we say. "Everything will be all right." When we think this way, we do not expect relationships to change and we are thrown off balance when they do. When we accept change as part of life, we know that our lives can and will change in relationships—for better and for worse. Like Whitehead and Hartshorne, Buddhists react against the concept of a substantial self, hard as rock, impermeable. Yet for process philosophy and for at least some Buddhists, there is a sense in which the self endures over time. The self is not something hard and substantial, yet it is a self. We know that in one sense we are each the same person throughout our lifetime. Even the Dalai Lama knows that he is the Dalai Lama. Is it the body that gives us this sense? I still have the same genetic makeup that I was born with. My father still recognizes me as his child. And I still recognize myself as Carol who was born to Janet and John, whose brothers were Brian and Kirk, and Alan who died in infancy. Yet the body too is constantly changing. I do not have the body I had at ten, at twenty, or at thirty. All the cells of our bodies, scientists say, renew themselves or change every seven years. So what is identity? Is it the mind or consciousness? But consciousness too is continually changing. I am and I am not the child I was at the age of three. Some of her experiences shape and haunt me to this day, but there are many others that are equally significant. We do have a sense of self, identity, and character, but this too changes as our experiences change. It is not rock hard. Hartshorne called this changing sense of self "nonstrict identity" in contrast to "strict identity," which admits no fundamental change.[13]

Does the process understanding that we are changed by our relationships mean we are only passive, only acted upon, only changed by others? This is what philosophers who asserted that the substantial self is like a rock feared. We can see that their fear was misplaced if we recall as common speech tells us that relationships are always a two-way street. We act and are acted upon. We are changed by others, we change them, and we have choices. Finding the black storks in the reservoir, I chose to stop to look at them. I chose to come back. I did not make them appear, but I allowed myself to be affected by the fact that they were there. They too had choices. The third time I came to see them, they resisted the urge to fly to the other end of the reservoir. This (as far as I can imagine the mind of a stork) was their choice. One stork did walk rapidly away from the others and me, but even it flew back after a while.

Native Americans have said, "We are the land."[14] Australian Aboriginals will sing certain songs only when they are evoked by a specific place in the landscape.[15] This suggests that some traditional peoples understand the boundaries of the self to be inclusive of particular landscapes or parts of the earth, along with the plants and animals that live in it. In other words, the relationship of the self to the land is also an internal relationship. You cannot take the self out of the environment it knows and expect it to be the same. When Native Americans were moved to reservations, they could not simply "get on with their lives." They had been uprooted from places that had deeply shaped their sense of self and community. Many whose livelihoods are or were tied to land still feel internally related to particular places. Some of those who moved to northern cities in the United States long for the sounds and smells of the south. Many Greek emigrants return home because they cannot live far from sea and salt air. Others of us wish to return to a place where we lived as children or young adults.

Some of us feel that we cannot move to another place, no matter what economic or personal reasons may call us. Others know that we must move from where we are or something vital in us will die.

Modern life is not structured to foster or take account of the inclusion of a sense of place within the boundaries of the self. We are expected to go away to college or move away from home in order to grow up and become independent. We are expected to move without difficulty to new places to find work or to get ahead. Nothing in our culture helps us to understand or to cope with the sense of "dis-location" we feel in a new place. "Don't worry. You'll adjust," we are told when we feel "home-sick." Nothing in our culture encourages us to mourn the loss of a place as much as we might mourn the loss of a loved one. Nothing in our culture helps us to understand that the wounding we feel is valid and deep. Nothing in our culture helps us to learn that the route to healing is developing connections to the particular landscape of a new place. Lack of relationship to a place may be one of the reasons that so many of us feel unhappy even when it seems that we have everything we need. Process philosophy did not elaborate the idea that the self is internally related to place, but this understanding is compatible with its notion of the relational self.

For process philosophy, "sympathy" is one of the main ways people deeply touch and change each other. The word "sympathy" comes from the Greek word "pathos," which means passion or deep feeling, and the prefix "sym" or with. Though sympathy is often said to mean "suffering with,"[16] it more accurately means "sharing deep feeling with" someone, because pathos includes suffering and love, joy and sorrow. Process philosophy asserts that deep feelings can be shared. We really can as we say "put ourselves into someone else's shoes" or "enter into someone else's soul." For those who believe in a substantial self or one that has only external relations with others, this is not possible. For process philosophy, it is. When a friend reads a book that helps her to make sense of her life, takes a vacation that makes her happy, or finds a job that is fulfilling to her, most of us rejoice. There may be a part of us that also feels jealous, or wishes we had the same luck, but this does not deny that we really do feel deeply the feeling of joy that our friend feels. When someone we love suffers, we also feel deeply with her. When my mother died, my friends were with me in my suffering. They perhaps did not know its exact quality, but they suffered with me nonetheless. Some of my friends said that in this way they were learning what it would be like when they lost their own mothers. When the mother of one of my friends died, my friend called me immediately, knowing that I would understand, not only because I had lost my mother but also because I knew her mother and the relationship they had. I was also changed as I helped my friend, feeling grateful that I had been able to give back some of what had been given to me. Sympathy is possible because the self is not rock-hard but permeable, able to be affected by others.

Those who begin with the notion that the self is substantial and independent try to explain our concern for others as "enlightened self-interest." From this point of view, birdwatchers would be concerned to save the wetlands for the birds because they want to come back to watch them. Even seemingly altruistic or generous actions are really motivated by long-term self-interest. Some religious ethicists also begin with the independent substantial self. They argue that ethical behavior requires "self-sacrifice." For them, the farmer must sacrifice his desire to build a hotel on the wetlands for the sake of a higher good, a world with birds in it. Process philosophy disagrees. Because we are internally related to each other, we can genuinely care about others as well as ourselves. A birdwatcher might work to save the wetlands for the birds, even if she knew she personally would never see another bird. The farmer might decide not to build out of concern for the birds because he understood they depend on fragile habitats. Process philosophy does not say that we always act out of sympathy for others. But it asserts that we can do so, and that sympathy is the source of ethical behavior.

It is not only personal relationships that can affect us deeply. When I wrote my thesis on the early writings of Elie Wiesel, I worried that I would not be able to understand the experience of a young man from Eastern Europe who was incarcerated in the concentration camps and lost his father there. Yet I lived with his descriptions of his experience for years. After Elie Wiesel read my dissertation, he said that he felt I really had understood his experience. Elie Wiesel's story affected me so deeply that I came to share his anger at and disappointment in the God of the Bible who had allowed his chosen people to suffer and be killed. Like Elie Wiesel, I would never again naively believe in the God of the Bible.

Sympathy is not limited to human others. Years ago when I used to cry at night because I did not have "a relationship," my little dog would climb into bed with me and whimper, then lick my tears away. To the full extent of the capacity of her dog consciousness, she sympathized with me. She may not have known the reason for my suffering, but she knew what suffering was. My suffering became her suffering and so she cried with me. When I became conscious of how she was suffering with me, I stopped crying to comfort her. The relationship we had was internal for both of us. Our characters were shaped by it. I too had saved her when she was suffering, terrified at having been left in the pound by her first owner.

Recently fledged birds sometimes help their parents feed the next group of nestlings. From an anthropomorphic and traditional Christian viewpoint, this behavior might be called altruistic or selfless. Or from a determinist sociobiological point of view, it might be said that the young birds' behavior is selfishly motivated to preserve their own gene pool, since the newer nestlings have basically the same genes as they do.[17] I suggest that there is another way to understand the behavior of the helpful little birds. If life is social through and through, then the birds were neither self-sacrificing nor selfish. They were simply cooperating in the enterprise of life. What if they felt sympathy for the cries of their hungry little siblings?

Sympathy also occurs between human beings and plants. For several years, I sat on my balcony nearly every day and looked at a very old pine tree in the garden next to mine. This tree became part of my internal landscape. I loved its greenness. I thought of how long it had been there, how many children had played under it, how many birds had nested in its branches. When a neighbor insisted that all the branches that leaned over his balcony be cut off, I felt as if a part of my own body had been violated. What the tree or the cells in the tree felt about my sympathy, I do not know, but I assume that it or they felt something.

This is not to say that sympathy is automatic. It may to some extent be instinctual, but it is also learned behavior. Monkeys raised without maternal care do not know how to treat their own infants. The same is true for human beings. In modern life, we are trained not to let sympathy become too strong. "This hurts me as much as it hurts you," my father said when I cried, "but it is my duty to spank you when you misbehave." "Don't flinch," boys say as they hit each other. "Don't let your feelings get in the way of your duty to fight and kill," soldiers are told. "Don't let your feelings get in the way of business," women executives are counseled. "This is no place for feelings," we are told. "Don't cry about it," we are told, "if you do, the others will lose respect for you." As we have seen, western tradition even asserted that God did not really suffer with the suffering of the world.

For process philosophy, sympathy—the ability to share the feelings of others—is fundamental and runs throughout the web of life. Process thinking asserts that the world is social through and through. We are constituted by our relationships and would be nothing without them. Relationships are not something we can choose to have or not have. Relationships are what we are. This is true of human beings and all other animals, and cells and atoms as well. It is also true of Goddess/God. There is no reality that is not social. Process philosophy looks at Plato's God, who is related to none, and asks how he could possibly become an object of worship. To put it simply, if God does not care about us, then why should we care about him? One can fear a lawgiver or a tyrant, but unless he shows us love or concern, we will not love or care about him. We will do what he tells us out of fear of punishment, nothing more. If such an unloved lawgiver or tyrant were to be killed, we would rejoice, not weep. Worship in contrast to fear implies that God is somehow worthy of our concern.

According to Hartshorne, the only Goddess/God worthy of worship is one who really does love us. In order to love, Goddess/God must also in some sense be personal. Hartshorne said, "A personal God is one who has social relations, really has them, and thus is constituted by relationships and hence is relative—in a sense not provided for by the traditional doctrine of a divine Substance wholly nonrelative toward the world."[18] This is a Goddess/God who can touch and be touched. Hartshorne's words come from a book he titled, tellingly, *The Divine Relativity: A Social Conception of God.* He offers a new interpretation of the philosophical contrasts "absolute" and "relative," disagreeing with those who state that God must be conceived as absolute and therefore not relative, not limited by or related to anything at all. Hartshorne asserts in contrast that Goddess/God is "surrelative," or "supremely related to all." As surrelative or supremely related, Goddess/God is intimately and perfectly related to all individuals in the universe. The divine sympathy is not limited to human beings. Goddess/God appreciates other animals, plants or the cells in them, cells, and even atoms and their particles with a degree of interest appropriate to them. Goddess/God feels our feelings and experiences our experiences and understands us as we would hope to be understood. Goddess/God sees and feels everything, not denying reality, but always with our best interests (and the interests of all others) at heart. Goddess/God really does rejoice in our joy, suffer with our suffering. Goddess/God will remember us and all other individuals for all eternity.

Human sympathy may fail, but divine sympathy never does. We are not always able to put ourselves in the other person's shoes. Our imagination is not sufficient, or our attention strays. Our own needs interfere some of the time, and other times there are too many other demands. Our perspective is not broad enough to take in the whole. This is inevitable because we are finite and limited. The divine sympathy is large enough to take in the entire universe and it is not disinterested. It is deeply interested in our struggles and achievements and those of all other individuals in the universe. In the Hebrew Bible the divine sympathy is expressed in the word "hesed," translated as "steadfast" or "overflowing love," and in the name "Immanuel," which means "God with us." In Christian tradition the divine sympathy is called love. Buddhists say compassion.

In order to explain the divine sympathy as surpassing all human sympathy, Hartshorne again introduces the principle of "dual transcendence." Goddess/God can be said to be absolute in one way and relative in another. We have seen that Goddess/God is relative in the sense of being fully and completely related to all. Goddess/God really is internally affected by what happens in the world. However, while human sympathy for others may waver, the sympathy of Goddess/God for the world is constant. Goddess/God is not fully present and fully related some days, but off in the clouds or sick in bed other days, as we are. Goddess/God is always related to all. In this one sense the divine sympathy can be said to be absolute and unchanging.

Insofar as the world is the body of Goddess/God, Goddess/God participates in our experiences from the inside. Goddess/God is "in" us in the sense of participating in our experiences. We are "in" Goddess/God insofar as we are part of Goddess/God's body. Because we are in Goddess/God's body, we can feel Goddess/God as a power within deepest self. Knowing Goddess/God as a power within the deepest self is not the same as saying that the deepest self is Goddess/God. All individuals in the world also have individual existence and degrees of creative freedom. Though we are connected and related, we are not one. Because you and I are not Goddess/God, we can enter into a personal or I-Thou relationship with Goddess/God. We can pray to Goddess/God for guidance. We can thank Goddess/God for sustaining the world and (for a time) our individual lives. And we can feel the presence of Goddess/God in our lives. The analogy Hartshorne suggests (admitting that is an imperfect one) is to cell in the body communicating with the mind.

There have been three times in my life when the divine sympathy and love was revealed to me with extraordinary clarity. Each of these was a point of crisis. At these times I received what seemed to me to be a revelation from the divine source. Twice it happened when I felt alone and abandoned. The other time was when my mother died. I have written about these experiences elsewhere, but as I recall them here, it strikes me that in the first two cases I received assurance of the divine sympathy and in the third of the divine love. The first occurred while I was writing my thesis on Elie Wiesel. For years I lived day and night with Elie Wiesel's anger at the God who had allowed the Holocaust to happen. During this time I was becoming more and more aware and more and more angry at the history and ongoing reality of injustice toward women. The God of the Bible, in whom I had deeply believed, seemed implicated in women's suffering. He had allowed himself to be known exclusively in male images, sent male prophets, and even a male savior. In his name women had been denied the rabbinate and the priesthood, had been told to obey their husbands, and had been burned as witches. The night I expressed my anger at God for all that he had done or let be done to women, I heard the words: "In God is a woman like yourself. She shares your suffering."[19] It was as if a voice spoke in my mind, but I had the distinct sensation that it was not my voice. It did not tell me that this female God or Goddess would come in glory to liberate me or other women, but simply that she shared our suffering. From a process perspective, the divine sympathy was revealed to me that night. Another night I was feeling personally abandoned, alone in the universe, without anyone who could understand me. Even Goddess seemed to have abandoned me. This time I expressed my anger at her. In the silence that followed, I again heard words that seemed not to be my own. The Goddess seemed to be saying to me: "The path you are on now will not be easy, but I will be with you all the way."[20] Again, from a process perspective, the divine sympathy was revealed. On the morning that my mother was dying in her own bed with my father and me beside her, I felt the room to be flooded with love.[21] This sensation was unlike anything I had ever experienced and unlike anything I might have imagined. It changed my life. Though it took a while for this knowledge to seep into my bones, I can truthfully say that from that moment to this I have understood that love really is the ground of all being. From a process perspective, this could be understood to have been a revelation of "love divine, all loves excelling."[22] Since then, the ordinary experiences of my daily life reveal the presence of this love—I feel it everywhere. In the first two of these experiences it was the divine sympathy—not the divine power to make things better—that was revealed. In the third instance as well, I did not receive the assurance that my mother would be granted eternal life—but rather that whatever happened to her or to me, love would never abandon either one of us.

Such experiences open me—as similar experiences have opened others—to the process view of Goddess/God as divine relativity, divine sympathy, and divine love. Yet it might be asked: Why focus on these experiences and not others? Aren't hatred and cruelty as real as love and understanding? These are important questions. Our insights and intuitions need to be reflected upon. Do they make sense of our whole lives and life as a whole? This will be the subject of the next chapter. For now, I invite the readers of this book to ask themselves whether they have also felt the divine sympathy and love in their lives. For without intuition or experience of the divine relativity, talk of it will probably fall on deaf ears.

As the experiences just described indicate, the divine sympathy does not offer to fix everything for us. Women (and others) continue to suffer. My life did not change overnight, though it has changed over time. My mother died. Looking back on it, I can see that one of the mistakes I made was to imagine that God or Goddess had the power to "fix" things in the world and in my personal life. What I understood through these three experiences is that there is nothing in life and nothing in death that can separate us from the divine sympathy and love.[23] The divine sympathy and love is known in and through our suffering, as well as in the overcoming of it. Goddess/God is not omnipotent, all-powerful, but Goddess/God is omnipresent, always here, a subject I will discuss further in the next chapter.

Does this mean that Goddess/God is merely a passive observer or passive recipient of our experiences? Goddess/God does not have the power attributed to an omnipotent God to fix everything or to destroy everything and start over. But this does not mean Goddess/God has no active power of any sort. When Goddess/God feels our feelings and experiences our experiences, Goddess/God makes a new creative synthesis of them simply by being present with us. We are never alone. The divine sympathy adds to and thus changes our experiences by appreciating them in the best possible way consistent with reality. In order to do this, Goddess/God makes a new creative synthesis of our experience by looking at it from a wider and more consistently loving perspective than we have. Goddess/God is internally and intimately related to every being in the universe, sympathizing with and remembering each of us in the best possible way.

Process philosophy says that Goddess/God influences the world through the power of "persuasion."[24] As persuader, Goddess/God takes account of the needs of each individual and the needs of the whole universe and encourages us to achieve our own ends in ways that will increase the beauty, harmony, and creativity of all beings in the web of life. Does this mean Goddess/God is continually speaking to us as I felt I was spoken to in two of the experiences I described? This may happen, but it is not part of most people's everyday experience, or mine. The world is the body of Goddess/God. From this we can deduce that Goddess/God's methods of communicating with us are myriad. Goddess/God can speak to us in the cry of child, the touch of a lover, the breath of the wind, the color of a sunset, the bark of a dog, the words of a book. To be open to Goddess/God is to be open to the whole world and its deepest meaning.

For process philosophy, touch most certainly is. Process philosophy views the world as social through and through and relationships as the building blocks of all life. Goddess/God is supremely related to all by a sympathy surpassing all other sympathies. The theological mistake known as God's unsympathetic goodness is overcome. Goodness is redefined as the ability to be fully and appropriately related. The importance of relationships in human and divine life has been neglected in many traditional philosophies and theologies. Could the reason for this neglect be that relationships, like the body and change, have been associated with women and with vulnerability, while the life of the mind and independence are reserved for men and God?

Chapter Four

Re-imagining Power

In a world where individuals other than Goddess/God really exist, the power of Goddess/God can never be "power over," but always and everywhere is "power with." Power over is domination. Power with is cooperation, partnership, and mutuality.[1] Goddess/God is the eternal Thou, fully and appropriately related to every individual in the universe. Goddess/God is the one for whom no individual in the universe ever becomes merely an "it."[2] The power of Goddess/God is to influence and be influenced by, to persuade and be persuaded. The power of Goddess/God is to suffer with and to enjoy all of life. The power of Goddess/God is to encourage all individuals to use their freedom and creativity to increase the beauty and harmony of the world. The power of Goddess/God is like that of a "mother, influencing, but sympathetic to and hence influenced by, her child and delighting in its growing creativity and freedom."[3] Creation is always co-creation. What happens in the world is never the result of any single will, not even that of Goddess/God. A multiplicity of wills has combined with chance to create the world as we know it. Because the world is co-created, everything that happens does not happen according to divine will. Because the world is co-created, everything that happens does not have a purpose in a divine plan.

This is a major change from traditional ways of understanding divine power in the West. In classical theologies and philosophies, the divine power is understood as omnipotence. Hartshorne correctly understood this to be a common theological mistake. "Omni-potence" literally means "all-power." Traditionally this is understood to mean God is all-powerful. If God has all the power, then everything happens according to the will of God, because no other individuals have any (significant) power. Other individuals are either doing what God permits or wills them to do, or they are doing things that the divine power can nullify or reverse whenever it chooses. Omnipotence can also be understood to mean "power over all." In other words, omnipotence is a form of "power over" that leaves no room for any power other than that of God. Omnipotence has been associated with God's dominion or rule over the world. It is more rarely noted that dominion is domination. Hartshorne made this connection clear when he said that to imagine the divine power as omnipotence is to imagine it on the model of a tyrant, a bully, a strong man, a king, or a feudal lord. He found it astonishing that theologians and philosophers almost never asked whether the power of the tyrant or the bully is indeed the highest imaginable form of power and therefore an appropriate model for understanding divine power. The notion of divine omnipotence gives rise to the well-known "problem of evil."

Job was one of the first to articulate questions that theologians and philosophers of religion have struggled with for centuries. "Why does a good man suffer while the wicked prosper?" Most of us have asked this question more than once in our lives. For Jews and some Christians, the problem of evil is focused on the Holocaust. How could a just or good or loving God have allowed six million people to die in the gas chambers? This problem is intensified for those who understand the Jews to be the chosen people of God. How could God have allowed his chosen people to perish in the flames of Auschwitz, Buchenwald, Treblinka? While writing my doctoral thesis on Elie Wiesel, I asked this question again and again. For many, the Holocaust has become a symbol for the spread of mass terror, dehumanization, and killing in our world. How could a good and loving or just God have allowed so many people to die at the hands of others in the twentieth and twenty-first centuries? I have struggled with these questions. There is a feminist question as well. How could a good and loving God have allowed women and children to be the poorest of the poor the world around? How could God have allowed black women, yellow women, brown women, and white women suffer at the hands of men in their own cultures and at the hands of conquerors and invaders? If God is good, why did he allow men in so many cultures around the globe to think it is their right always to dominate and sometimes to abuse, rape, and even murder women?

The assumption hidden behind all of these questions is that God has the power to stop suffering and evil. When God's power is conceived as omnipotence, the problem of evil is formulated thus: If God is all-powerful, why does he not intervene to stop the suffering of individuals and groups? Or to put it another way: Why does God not reward the good and punish the wicked? Theologians then propose several (in my opinion unsatisfactory) answers to these questions.

One answer is that the good will be rewarded and the wicked punished at some point in time or space. In an ending added to the book of Job by an ancient editor, Job's honor as well as his family and fields were restored. In traditional Christianity, the good will be rewarded and the wicked punished, if not in this world, then in the next. The familiar pictures of Saint Peter's gate and heaven and hell express this understanding. Others assert that the good will be rewarded and the wicked punished at the end of time, when the Messiah comes in glory to establish his kingdom on earth. The Hindu theory of karmic reincarnation is yet another way of explaining that the good will be rewarded and the evil punished. For Hindus this will not happen in this life but in our next incarnations. In popular understanding, the wicked will come back as animals, women, or untouchables, while the good will come back as rich men or saints. Good women will come back as men. One of the attractions of Hinduism is that its doctrine of reincarnation can be understood to mean that all suffering has a purpose in that it leads eventually to enlightenment. Traditional Hinduism and Christianity and some forms of Judaism agree that there is some time when the good will be rewarded and the wicked punished.

The other answer to the problem of evil is to say that God's ways are not our ways. This is one reading of the original ending of the book of Job. In it, God speaks to Job, asking if Job was there when God created the Leviathan, a monster of the deep, and the hippopotamus. A God with the power to create Leviathan was understood to be an all-powerful or omnipotent God who should not be held to our human standards of right and wrong. What seems to be evil or unfair to us with our limited perspectives may not be so in the higher and grander scheme of things that has been set forth by the almighty God. Read this way, God's answer to Job is an assertion of power. It comes down to this: I (God) have all the power, and who are you to question me? The notion that we have no right to question the divine power keeps many of us from asking questions in our hearts. Yet, the sheer assertion of unfathomable divine power is not something that all of us can feel comfortable with. Therefore, it is usually added that suffering must and indeed does have some purpose in the divine plan. In this way God's unfathomable power is rationalized. We need to suffer to learn a lesson, in order to learn to be more compassionate to others, and so on. The second explanation leads back to the first.

For many people today, myself included, these traditional answers to traditional questions are not satisfying. Many Jews have said that it is obscene for anyone to say that the suffering of the Jews in the concentration camps serves a higher purpose. No conceivable reason or purpose could possibly justify the deliberate humiliation and degradation, followed by death, suffered by so many millions. Nor could reward in any other world make up for what millions of Jews suffered in this world. Any God who could have stopped the Holocaust should have done it. Many sensitive people extend the indictment. Any God who could have stopped the spread of AIDS in Africa should have done it. Any God who could have fed hungry women and children should have done it. Any God who could have stopped fathers in every country in the world from raping the bodies and souls of their own infant daughters should have done it. For those asking these questions, it may seem like no answer at all to reassert the formulaic phrase: "His ways are not our ways." For many people it comes down to this: Either God does not exist, or God is not good and just and therefore not worthy of our love, concern, and worship.

Process philosophy asserts that all of these questions and answers, though heartfelt, are misplaced. The "problem of evil" has been wrongly formulated because it is based on the wrong premise. Process philosophy offers another way of looking at the questions and the answers. From traditional religious and traditional atheistic perspectives, it is a startling one: omnipotence is not the kind of power Goddess/God has or should have. Goddess/God did not create Job's suffering. Goddess/God did not create the Holocaust. Goddess/God did not cause the spread of AIDS in Africa. Goddess/God did not create the situations in which mothers and children go hungry. Goddess/God did not cause fathers to rape their daughters. Goddess/God did not create the suffering in the world, and Goddess/God cannot alleviate it without our help. The power of Goddess/God is not power over, the power of a dominator, but power with, the power of cooperation and co-creation.

A common twentieth-century interpretation of God's appearance to Job is compatible with process understanding. When God spoke to Job, a number of biblical scholars suggested, God did not offer any explanation or justification for Job's suffering. What Job learned from his encounter with God was not that his suffering had a meaning, or that God was all-powerful, but that only and significantly that God was there. God had been present at the creation of the world and he was present with the suffering of Job.[4] This reading of Job is fully consistent with process philosophy. In my opinion it the only rationally and religiously satisfying answer to the problem of evil. Goddess/God is not omnipotent (all-powerful), but Goddess/God is omnipresent (always there). The presence of Goddess/God is the divine sympathy discussed in the previous chapter.

"Sounds good," you might be thinking, "but if Goddess/God is sympathetically related to all individuals in the universe and influencing them toward the good, then how do evil and suffering get going in the first place? And how have they gotten so out of control?" Process philosophy offers two answers to these questions: chance and choice. Choice is part of one traditional answer to the problem of evil as well. It is important to distinguish the process understanding of choice from this traditional one. God, some theologians have said, first created the world and then gave freedom of choice to human beings. Being omnipotent, God can intervene in the world at any time, but he does not, because he wants to leave space for human freedom. God (so to speak) steps back from controlling everything in order to give us room to act. This leaves open the possibility that God could step in to set things straight if he chose to do so. Evil is created by human beings, with the permission of God. Does this view make sense? Good parents step back to allow their children to make choices. But no good parent would allow an eight-year-old to torture and behead kittens. God, on the other hand, seems to be quite willing to let us torture and behead each other. If God is something like a good parent, then God ought to intervene when we go too far. At this point the two standard responses to the problem of evil are repeated. Those who are tortured and beheaded in this life will be rewarded in some time or place. Or who are we to ask, because God's purposes are unfathomable. The traditional view that evil is a result of human choice cannot explain evil that is not a result of human choice—for example, the spread of disease or the destruction caused by floods and hurricanes. Some theologians therefore invented the category of "natural evil" which is said to include death, disease, and natural phenomena such as hurricanes, floods, and earthquakes. The degree to which human beings have altered nature with modern technologies has blurred the distinction between human and natural evil. Are devastating tornadoes now being experienced a result of natural processes or the result of global warming? If a river floods, is it because human beings have filled in the wetlands where the water collected? In classical theism the question of why God allows so-called natural evil to exist is variously explained (or not explained).

Process philosophy rejects this whole line of thinking. In the process view it makes no sense to blame Goddess/God for things that are either inherent in the nature of the world or caused by individuals other than Goddess/God. For process philosophy, death (for all individuals other than Goddess/God) is inherent in the nature of the world. I will reflect on this in chapter five. Process philosophy argues that most of the other things we experience as suffering or evil are created by choice and chance. Because process philosophy views the world as genuinely co-created, the role played by choice in the creation of evil and suffering is understood quite differently than it is in many traditional theologies and philosophies. Chance can play no role in traditional thinking that asserts that everything happens for a purpose or according to the will of God.

In contrast, in process thinking, creative freedom is one of the defining characteristics of reality. Goddess/God did not create the world out of nothing and then (as a kind of afterthought) decide to give freedom to some of the creatures. Freedom has always existed in Goddess/God and in the individuals (at one time perhaps only particles whirling in space) that make up the body of Goddess/God. Creative freedom is understood to be inherent in the nature of all reality. As Hartshorne said, Goddess/God "makes things make themselves."[5] The exercise of creative freedom is always relational, as individuals always make themselves and the world in relation to the creative freedom of other individuals. Creation is always co-creation.

Choice is the exercise of creative freedom. All individuals exercise creative freedom as they go about finding a niche for themselves in relation to other individuals in the web of life. To a great extent, individuals must co-operate in order to survive. Birds that nest in the pine tree outside my window carry its seeds to fertile ground. Moss protects its bark from wind and sun. A cat rests in the shade of the tree while the remains of a mouse it has just eaten begin to disintegrate into the soil providing nourishment for the tree. Conflict and suffering inevitably result from the fact that there is more than one individual in the universe. Individuals compete for territory, mates, and food. However, most animals other than humans kill only to eat. And as Hartshorne has noted, for individuals other than human beings (and those affected by human beings), suffering is usually short and death swift. For millions of years, the lives of human individuals did not differ greatly in these respects from the lives of other individuals in the universe. The symbol systems of many indigenous peoples encode an acceptance of death and an understanding of the need for all beings to cooperate in the web of life.

The really significant suffering and evil in the world is that created by human beings who have constructed technologies including tools and weapons that allow us to manipulate the environment and the lives of other individuals in significant ways. Human beings have caused a great deal of suffering for ourselves and other life forms through the exercise of creative freedom or choice. God is not to blame for the suffering human beings have created, we are. Human choice is real and it has consequences. But it is more complicated than that. Sometimes we recognize the consequences of our actions and sometimes we do not. Sometimes we think we have done the right thing and only later realize that we have not. Moreover, when the results of human choices become encoded in social structures and cultural symbols, they take on a life of their own. Most of us do not choose to be racists or sexists, or to cause other human beings to go hungry or species to become extinct. Yet we participate in social and cultural systems that have these effects.

In contrast to traditional religious thinking in which everything that happens is attributed to the will of God, process philosophy asserts that chance also plays an important role in the universe. Many things happen in the world that were not strictly speaking intended by any individual and certainly not by Goddess/God. How does this come about? Let us say that I decide to do one thing and you decide to do another and that both of these seemed to be good choices at the time. Yet the combined result of our two choices cannot be foreseen by either of us. You did not intend it, I did not intend it, but there it is, and it has hurt or helped someone. For example, I decide to drive to the store to get milk for my family. At the moment I round the corner, you decide to run into the street after you dog. The car I am driving hits you and we are both hurt. This was not intended by you or me or Goddess/God. It just happened. It was chance. Not all things that happen by chance are bad. You and I may meet in the hospital and become friends or we may decide to work together to make a blind corner safer for others. What we do with our chance meeting is choice, but the fact that we met due to an unintended accident was chance. Let us imagine that one of us was killed in the accident. It would do our friends and families no good at all to search for a meaning or purpose in what happened, nor is it appropriate for them to comfort each other by saying, "It must have been the will of God."[6]

Chance occurs throughout the universe, and it is one of the ways the evolutionary process works. Of all the seeds blown by the wind, one drops in a place where it can take root and over time a species exists in a new place. Another seed, a mutant, also falls in an inviting place. Over time a new kind or color of plant may develop in that place. The new plant may begin to displace the plants that had once thrived in the place, yet it may also attract other plants and animals that can live in cooperation with it. When we understand that seemingly trivial choices (even that of one seed or of its cells) may have enormous consequences, we understand that choice is more complicated than we might have thought. Because we are finite and limited and interdependent, our choices have consequences that we cannot imagine.

Because choice and chance exist throughout the universe, the power of Goddess/God to control and shape reality can never be understood to be absolute or unlimited. Is a less than omnipotent Goddess/God worthy of worship? Haven't we always worshipped God because we imagined him to be powerful enough to control the world? That these questions arise again in my mind (and perhaps in yours) is testimony to the power of traditional conceptions of divine power. Process philosophy asserts that the power of Goddess/God to control the world cannot be absolute and unlimited if individuals in the world really exist. If we exist, we must have some power of life, some power to create, some power of choice. Otherwise we would only be aspects of the divine life. According to process philosophy, if God has all the power, then there is none left over for us.

Traditional theology and philosophy tried to have it both ways. Omnipotence and human freedom. They do not really go together. Trying to imagine that they did created problems for thinking individuals who asked: Are we really free if our freedom can always be cancelled by an all-powerful God? Only through a kind of double-speak can traditional theology affirm that human freedom can be understood together with divine omnipotence. Moreover, the notion of an omnipotent God who rules over a world filled with suffering leads to the lingering suspicion that God might be a sadist or at best indifferent to the sufferings of the world.

For those struggling with questions raised by traditional theism, the notion that "all is One" found in some forms of mysticism and many forms of traditional Hinduism[7] may seem appealing. Philosophically, this view is called "monism," which means "one-ism" or "oneness." Its basis in experience is a mystical feeling of being part of everything with no separation at all. Monism rejects traditional theism's view that an all-powerful God stands outside the world, ruling and judging it. Monists insist that the divine power is present in all of reality. All is God and God is all. In traditional Hinduism it is said that the world is "maya," or illusion. In reality we are so fully part of the unfathomable Oneness that it makes sense to say that our distinct individualities and our individual joys and sorrows are illusory. Taking this view to an extreme means that I am not really typing these words on a computer created by human minds and hands. In reality the One is typing them on a computer created by the One.

In asserting that all is One, monism offers its own solution to the problem of evil. If all is One, then good and evil are also One. The One is anger and reconciliation, war and peace, hatred and love, death and life. All that exists is part of the cosmic dance in which the dancer is One. The One is the father reading his child a story, the child being read to, and the author of the story. The One is the father raping a child and the child being raped. Individuality is an illusion, maya. Suffering is also an illusion, maya. Our sense of outrage at a father raping a child is illusion, maya. Our struggles to protect children too are illusion, maya. There is only One. All is One.

Process philosophy offers a different view. For the Goddess/God of process philosophy, the rape of a child is no illusion. It will exist for all eternity in the divine memory. Goddess/God's suffering with a raped child is immeasurable. Goddess/God's outrage at every violation of life is also immeasurable. With all the power Goddess/God has, Goddess/God is working to persuade every single one of us not to violate innocent children and to change structures that allow this to happen. Goddess/God also suffers with the father who rapes his child and (with a wider sympathy than most of us could ever have) sympathizes with the traumas he must have suffered in his own life. But this divine sympathy by no means makes Goddess/God indifferent to the rape.

While affirming the mystic's insight into the intimate connection of all beings in the web of life and the presence of divine power within all of reality, process philosophy also asserts the reality and importance of finite and temporal life. For this to be so, there must be more than one dancer in the dance of life, more than one actor in the cosmic drama, more than one musician in the symphony of the universe. Some power to affect others and the world is and must be ours. Some power is and must belong to all individuals in the universe down to the particles of an atom. For process philosophy, change and touch and with them creative freedom are the hallmarks of life in every individual in the universe from the smallest particle to Goddess/God. Each individual in the world has a degree of creative freedom appropriate to it. We are not simply acting out parts in a dance, symphony, or drama written by Goddess/God.

There are real individuals in the universe. Individuality is not rock-hard (or strict), but neither is it an illusion. We co-create the dance, drama, and symphony of life. We have a degree of choice in our lives. Because we co-create the world, Goddess/God cannot be omnipotent or all-powerful as asserted in traditional theism. If Goddess/God were all-powerful, our freedom would be an illusion. Nor can all be One, for in this case our very existence is an illusion. If our existence and freedom are real, then the power of Goddess/God cannot be unlimited, cannot be all there is. In distinction from both traditional theism and traditional monism, process philosophy asserts that individuals other than God have some degree of freedom. As free and creative, all individuals have some ability to affect each other and the course of life as a whole. Because the world really is co-created, the adventure of life involves genuine risk in all its stages for all its participants, including Goddess/God. Goddess/God alone does not create or control the fate of the world or the fates of the individuals in it.

But, it might be asked again: Is a Goddess/God who is not in control of everything worth worshipping? Hartshorne answers that the only divine power both capable of being worshipped and worthy of being worshipped is one that is not omnipotent.[8] A

God with power over the world might be held in awe, feared, and even obeyed, but the lingering suspicion that he is a tyrant or a sadist makes worship inappropriate. Moreover, if God has all the power, we cannot worship him, because we do not really exist. Our worship would be a manifestation of the love of the God for him- or herself. In contrast, the Goddess/God of process philosophy can be worshipped by individuals who really exist. And the Goddess/God of process philosophy is worthy of worship because Goddess/God cares deeply about all of our individual lives and the fate of the world as a whole.

"But," you might be thinking, "what good is this care and concern if Goddess/God does not have the power to end our suffering? Hasn't belief in God always been about finding a way to end suffering?" It is true that the Goddess/God of process philosophy does not offer the comforts offered by many forms of traditional religion. There is no heaven or hell—except for the ones we have created here on earth. None of us will be offered another incarnation in which our sufferings will be minimized or in which we can learn the lessons we failed to learn in this one. Human life is short and the lives of many other individuals in the universe are shorter. This is all there is. The world into which we were born was already shaped by choices that neither you nor I nor Goddess/God created. We must make the best of it, for this is all we have. I find the process view appealing because it is realistic. It does not deny or gloss over the reality of suffering in our lives and in our world. It affirms our outrage at suffering and our desire to enjoy life. It places the responsibility for creating a better world firmly in our hands.

Though the power of Goddess/God is exercised in cooperation with all of the other individuals in the world, the power of Goddess/God is greater than that of any other individual in the world. What then is the power of Goddess/God if it is not omnipotence? Hartshorne envisions the power of Goddess/God in two ways, according to the principle of dual transcendence. I will call these the power of Goddess/God as the ground of all being and becoming[9] and the power of Goddess/God as sympathetic and persuasive influence. The power of Goddess/God as the ground of all being and becoming is the unchanging or eternal aspect of divine power. It is the foundation of the changing universe. As the ground of all being and becoming, Goddess/God is the source of cosmic order. Another way of putting this is to say that Goddess/God sets "limits to the disorder inherent in freedom."[10] What we know as observed tendencies or patterns within nature conform to values or principles of being and becoming that exist as a potential within Goddess/God. Exactly what these are is difficult to determine from our fragmentary perspectives. Hartshorne thought that these must include a set of physical laws that govern the behavior of the basic elements, as noted in chapter two.[11] From a process point of view we might suggest that the fundamental principles of being and becoming also include individuality, interdependence or relationship, feeling, sympathy or love, creative freedom, and enjoyment. As the ground of being and becoming, the divine power keeps the balance of any possible universe tipped (however slightly) in favor of life and creativity, thus preventing it from disintegrating into sheer chaos. For Hartshorne the world is the body of Goddess/God. Thus we can say that the divine power as the ground of being and becoming holds the world that is the divine body together. We can speak of the fundamental principles that undergird the cosmic order as reflective of the love of Goddess/God for the world. The power that grounds and sustains the process of being and becoming is absolute and unchanging and can be seen as constituting the nature or character of Goddess/God. This means that Goddess/God cannot decide some day to stop the creative process, nor can Goddess/God decide to not to be sympathetic and loving. However, the power of being and becoming as absolute and unchanging power is abstract until it is made actual or concrete in relation to individuals in some possible world.[12]

In relation to any actual world, Goddess/God is relative and changing. In relation to the world, the power of Goddess/God is power with and not, as traditionally imagined, power over. In relation to the world, the power of Goddess/God is sympathetic and persuasive, as we have seen in the previous chapter. Goddess/God is directly and intimately involved in the lives of all individuals that make up the divine body, feeling their feelings and responding to them appropriately. The power of Goddess/God is infinite in the sense that it is the power to be with every individual in the universe in every moment. This will not change. But in relation to individuals in any possible world, the divine power is relative and does change, because Goddess/God really does enter into sympathetic relationships, feeling the feelings of changing individuals. As we have seen, the divine sympathy is not only passive, which is to say moved by the experiences of others. The divine sympathy is also active, responding appropriately to the feelings it feels, changing the experiences of every individual in the world by participating in them. This is the persuasive power of Goddess/God.

According to one English translation of the book of Job, after God spoke to him, Job "repented."[13] Traditionally this has been understood to mean that Job acknowledged that he was wrong to question a higher power against which he could never win. However, another interpretation is possible. The word often translated as "repent" in this passage is a form of a Hebrew word that which means "to turn." It has been suggested (and this reading is consistent with a process understanding of the divine sympathy) that Job understood that God had not abandoned him in his suffering. In this sense Job did not "repent" of having questioned God. It was in and through Job's questioning that God revealed his presence. Understanding that God was present in all his experiences, the bad as well as the good, caused Job's heart to "turn," transforming Job's experience of his suffering and his anger at God. Though Job's anger at God was transformed, it was only in and through his anger that he came to understand God's presence in his life. Job's anger was an important part of his process.

The persuasive power of Goddess/God cannot be coercive because it is expressed in relation to free individuals. Yet when we experience the presence of a sympathetic divine power in our lives, our hearts, like Job's heart, really can "turn." We know ourselves to be understood. This may enable us to open ourselves to feeling the feelings of other individuals. As we share the joy of others, our joy is increased. As we share the suffering of others, we become inspired to help those who are suffering and to change the world. "But," you might be thinking, "if the divine power works this way, then it isn't working very well. Suffering seems to be on the increase in our world." This point is well taken. If the divine power exists and is trying to persuade all of us to open our hearts to others, it does not seem to be doing a very good job. There are far too many individuals in our world—including you and me at least some of the time—whose hearts are not open to feeling the feelings of others. Or whose hearts are open only to feeling the feelings of their own friends and family or the feelings of some others in their own ethnic group, religion, or nation. Structures and symbols of domination are deeply embedded in societies and cultures. If the divine sympathy is influencing us to open our hearts to the world, it is clear from experience and observation that we as individuals and as groups are free to listen or not listen to the persuasion that is being offered to us. If Goddess/God is "She [or He] who hears the cries of the world,"[14] the human race as a whole is not doing a very good job of creating itself in the image of Goddess/God. The tragedy of human life is that so many of us have used our freedom for so long with so little concern for the joy and suffering of others. Whether human-inflicted suffering increases or decreases in our lifetimes is to some extent dependent upon the choices made by each one of us. The divine power is active but not omnipotent. The world will be changed for the better only when free individuals open their hearts and with Goddess/God co-create a different world. This is the subject of chapter seven.

At this point you might be asking another question: If Goddess/God knew that in a world with genuine freedom there could be so much suffering, wouldn't She or He have been better off not creating the world at all? One answer to this question is that if individuals really are free, then Goddess/God does not know the future. Goddess/God might be able to predict (on the basis of greater experience and perspective than you or I have) that this or that might happen, but Goddess/God cannot know what will happen until it actually does happen. Goddess/God is involved in the genuine risk of freedom. In addition, process philosophy suggests that there probably never was a time before creation when Goddess/God could have entertained the thought of whether to create it or not. For process philosophy it is not possible to think of Goddess/God apart from the relationship to a world that constitutes Goddess/God's body and very being.[15] Thus we cannot imagine a point when Goddess/God could have examined the possibilities inherent in a world with genuine freedom and decided to create it or not create it. If we could imagine a time before creation, it might be appropriate to ask: Was it worth the risk? But we are not in position to imagine such a time, because the only Goddess/God we know is the Goddess/God whose body is the world. The question we can ask (of ourselves and other human beings) is: Would you rather be alive or not? We know that the answer to this question is, in the vast majority of (but not all) cases: I would rather be alive. This is a not insignificant fact.

Another question raised by the notion of a Goddess/God that is not omnipotent is: If Goddess/God is not all-powerful, what does it mean to say that Goddess/God sustains the world when we all know that the earth and its human and non-human creatures are being destroyed on a daily basis? When the possibility of even greater devastation in a nuclear war is a cloud that hangs over all that we do? Is a Goddess/God who may not have the power to save this earth worthy of worship? In the 1980s while working on a slide show about suffering in war and the threat of nuclear annihilation, I was haunted by the thought that the earth could be destroyed. I read that only insects and grasses would survive a nuclear war.[16] I imagined an even worse disaster. How could Goddess/God have created the earth only to allow us to destroy it? I asked this question again and again for years. Hartshorne asked this question as well.[17] The answer that finally came to me was this. If most life on this earth is destroyed, Goddess/God would suffer immeasurably in both body and soul—infinitely more than I suffer when I imagine this possibility. Such an outcome would in no way be the will of Goddess/God. It would be caused by the combined wills of other individual beings, most of them human. But were this to happen, Goddess/God would continue to create with whatever bits or particles of life are left. For me this insight means that we who work to end suffering and save the environment of the earth do so without assurance that the earth abundant and thriving with life can be saved, a point I will return to in chapter seven. There is a possibility that the adventure of life on earth will be ended or greatly curtailed. Goddess/God is working with us to avert this possibility, but Goddess/God alone does not have the power to stop the destructive forces that human beings have set in motion. The world really is co-created. The fate of the earth is in our hands.

Most North Americans (and some Europeans) take freedom for granted. Some of us even imagine that it is nearly unlimited. "Anyone born in America can become president," we are told. "If you work hard enough, you will get ahead." "Raise yourself up by your own bootstraps." "If we think abundance, we will create it." "Some day we will colonize the moon." "Modern science will discover a cure for every illness." Others are quite clearly aware that freedom is limited. "I wish I knew how it would feel to be free," Nina Simone sang, protesting the fact that black people are denied the freedom other Americans take for granted. "I am a woman struggling to be free," my Greek friend Axiothea says, clearly aware that this is no easy task in a traditional Greek village where women could scarcely venture outside their homes unescorted twenty years ago.

While not in any way condoning historic injustices, process philosophy can help us to understand that freedom always involves limitations. No individual, not even Goddess/God, can do absolutely anything he or she wants to do. We live in bodies, in space and time, and in relationship to each other. Process philosophy can help those privileged by race, sex, class, or history to understand that the freedom they take for granted is not absolute but conditional. For example, the achievements of elite white males are not simply a result of the exercise of creative freedom, for rich white men are aided and abetted by a system in which they are programmed to achieve and assisted to do so. On the other hand, process philosophy can help those struggling to be (more) free to understand that there is a degree of creative freedom in every situation. This means that it is possible to create change (however small) in the world into which we were born. No one creates out of nothing, but we all can and do create. No one's freedom is unlimited, but we all do have a degree of freedom. If we listen to the persuasions offered by Goddess/God and other individuals, we can make changes in our individual lives and we can work together to change the structures of inequality and oppression.

But what, you might be thinking, does this view of a Goddess/God that is sympathetic but not omnipotent offer that atheism or humanism does not? Don't humanists and atheists also understand that this world is the only world and that we are the only ones who can change it for the better? This is a good question. Process philosophy offers an understanding of the web of life quite different from that of many humanists and atheists. Many humanists have what has been called an anthropocentric or human-centered view. When they rejected God, most "hard-headed" thinkers (this is a phrase they use of themselves) also rejected the notion that the whole universe is alive and that we are joined with all people and all beings in a web of interconnection. For all their flaws, traditional religions often foster a sense of interconnection within religious communities, sometimes among all human beings, and even with the whole of creation. Many of the problems we face today stem from placing undue confidence in "man's" ability to "shape" the world through scientific knowledge and technology. Process philosophy can offer humanists a more holistic view of the world and an understanding of the interconnections of all individuals in the web of life.

But couldn't the holistic view be adopted without adding Goddess/God to the mix? Of course it could, and many humanists and atheists are beginning to adopt a more holistic view in face of the ecological crisis and the suffering created by human arrogance. So what does Goddess/God add? In one sense nothing, in another sense everything. Humanists, atheists, and those who believe in a non-omnipotent Goddess/God will probably be asking many of the same questions about how to achieve greater peace and justice in our world and will probably be active in many of the same causes. Yet for those who sense the presence of Goddess/God in all their relationships and activities, the world is transformed. When I rejoice in the joy of the great tit hopping around outside my window, I know Goddess/God is rejoicing too. When I speak to others about my hope of achieving peace with justice rather than retaliating in violence, I feel that to the best of my limited knowledge I am creatively responding to the persuasions of Goddess/God. When I suffer, I know that Goddess/God has not abandoned me. When I witness the suffering of others, I know that Goddess/God is suffering too. I also know that Goddess/God shares my outrage at injustice. I experience Goddess/God as persuading me and others to respond creatively—to help those who suffer and to try to create a world where human-inflicted suffering is lessened. A small difference perhaps—or all the difference in the world.

For process philosophy, the power of Goddess/God is not power over, but power with. The world is supported and sustained by Goddess/God, but it is co-created by Goddess/God and all other individuals. Everything that happens in the world is not caused by Goddess/God. Chance and choice play a role. Belief in divine omnipotence is a common theological mistake. Yet it is rarely recognized as such. Why have we not recognized the degree to which conceptions of divine power as power over are modeled on images of the strong man, the king, the feudal lord, the bully, and the tyrant? Why is it that we do not readily imagine divine power on the model of the "mother, influencing, but sympathetic to and hence influenced by, her child and delighting in its growing creativity and freedom?"[18] Could it be that our inability to imagine a divine power that is related to the world but not dominant over it also stems from a rejection of the female body that is deeply embedded in our culture?

CHAPTER FIVE

LIFE IS MEANT TO BE ENJOYED

For process philosophy, the world is the body of Goddess/God. Goddess/God does not reside in heaven, nor is heaven our goal. The world is our true home. This life is meant to be enjoyed. There is no other. To enjoy life is to cherish the beauty of each living thing, to be interested in diversity and difference in the web of life. To enjoy life is to delight in the pleasures of the body. To enjoy life is to find meaning in relationships, not only with human beings but also with other individuals in the world, and with Goddess/God. To enjoy life is to influence and to be influenced by others. To enjoy life is not to have power over, nor to have it all, but to share. Process philosophy does not idealize life. The life that is to be enjoyed is not without suffering. In a world with many individuals, conflicts of interests are inevitable. In a thoroughly relational and social world, we are born into a web of pre-existing patterns and structures of justice and injustice that shape and limit the possibilities of enjoyment. We will not escape death. Still, this life is meant to be enjoyed. Goddess/God does not exist in a realm apart from joy and suffering. Our joys and our sufferings are shared, responded to, and become part of the life of

Goddess/God. The more we enjoy life and increase the joy of others, the more we contribute to the life of Goddess/God.

Enjoyment, like happiness, has two distinct meanings and process philosophy intends them both. On the one hand we speak of direct momentary joy or happiness as contrasted to direct momentary pain or sadness. When process philosophy says that life is to be enjoyed, it means that life is meant to be filled with moments of joy in this sense. Good food, good friends, satisfying work, walking in the woods, flowers in the fields, dancing until dawn, the sea around us, the stars overhead, a lover's kiss. Yet we can also speak of enjoying a life that includes pain and sadness. In this sense our enjoyment is reflective. Looking back over the lives we have lived so far, most of us would say that we have enjoyed living. At minimum most of us would say that we would rather be alive than dead. This means that for most of us moments of joy outweigh moments of pain in number, intensity, or value. Only thus can we truthfully say that we enjoy living. This is why we go on. Or at least, process philosophy would say, we should be able to say that we enjoy life. If we cannot, we ought to ask what is wrong. Have we been so harmed by others that we cannot enjoy life? Have we been taught that we have no right to joy?

All well and good, you might be thinking, for those who live long and do not suffer greatly. But what about those who die young or suffer intensely in a final illness? What about those who live in poverty or conditions of war? What about all the starving children in Africa or India? Surprisingly, process philosophy states that the purpose of all of their lives is enjoyment. For process philosophy no one is suffering for Goddess or God, for sins or karmic debts, or for the greater good. Those who die young have short lives. There is no evidence that the quality of their enjoyment is diminished by not having lived longer. Suffering in final illness does not negate the enjoyment that has earlier been experienced. Moreover, if death is accepted as part of life, the time of dying can also be enjoyed, as will be discussed shortly. The fact that some children are starving is due primarily to habits and patterns of behavior caused by human greed, humanly caused warfare, and human beings' lack of concern for the environment. Such suffering is the last thing Goddess/God would have wanted. Goddess/God wants us all to enjoy life. Where suffering has a human cause, we need to take responsibility for our roles in creating it and do what we can to change it, for those who suffer in this life will not get another chance.

Those who are not poor and not living in war-torn lands often imagine that we enjoy life while others do not. "After all," we say, "we have everything, and they have nothing." Yet it is not necessarily true that those with material comfort enjoy life more. Many of those with economic and social privileges are not happy, while others have great capacity for joy even in the midst of poverty and violence. My friend Ellie, who was born in the Greek village where I live, has had a difficult life. Her father was a refugee from Asia Minor and died of tuberculosis when she was three years old. Her baby sister was given up for adoption and was not seen or heard from for over fifty years. Though she had dreamed of becoming a lawyer like her father, Ellie was lucky to have finished sixth grade. Her grandmother urged her to sell the eggs that would have been breakfast to buy schoolbooks. Ellie's mother and stepfather arranged her marriage when she was fifteen years old, against her will to a man she did not love. Soon thereafter, the Nazis occupied Greece. Many people were killed. Many others starved to death, including the grandmother who had loved her and taken care of her. The Germans took everything of value from her home. At the end of the war, Ellie became pregnant, but before she gave birth, her husband was put into prison where he was beaten and tortured for seven years because he had refused to sign a document vowing that he was not a Communist. Ellie was forced to live with relatives who did not love her. When her husband came home, he beat the daughter he had never known. He was often at sea for months at a time, while she was left alone with three children. In those days women could not even walk freely on the streets. And yet, when I met her, Ellie loved life far more than

I did. She loves going out. She loves the sunshine. She loves the sunset. She loves cats. She loves her children, grandchildren, and great-grandchildren. She loves me. She loves to cook mousaka and squash pies. She is interested in the lives of everyone in the village. She will sing or dance at the drop of a hat, even though she is now over seventy-five years old. I have learned more from her and other older Greek villagers about the meaning of life than I did in all my years of study.

Alice Walker's character Shug was speaking about enjoying life when she told her friend Celie that "it pisses God off if you walk by the color purple in a field somewhere and don't notice it."[1] In encouraging Celie to enjoy the beauty around her, Shug did not deny Celie's suffering. Shug knew that Celie had been raped by her father and was being beaten by her husband. She knew that Celie was poor and black and considered ugly. Yet in helping Celie to understand that there was beauty even in her meager existence, Shug handed her a lifeline. As Celie learned to enjoy the fields flowering in the color purple and her own growing feelings for Shug, she began to see that there was a way out of her misery.

To recognize that those who are suffering have capacity for joy does not mean that there is no need to change structures that cause unnecessary suffering. Celie obviously enjoyed life more when her husband stopped beating her and she would have enjoyed it even more if she had not lived in a world shaped by racism and sexism. When we understand that life is meant to be enjoyed, we can see ever more clearly the wrongness of any humanly created system that limits the opportunities of some individuals—human or nonhuman—to experience joy. In our world the greatest obstacle to enjoying life is created by structures of injustice to other human beings and the natural world that shape our lives. While it is possible to enjoy life under conditions of poverty and violence, it has never been easy and it is becoming more difficult. In the past poor people could at least count on finding fields flowering in the color purple. Yet as global capitalism cuts down forests and introduces chemicals into the water table, much of the world is becoming poisoned and turned to desert. When this happens men leave home to seek work, and women's days are shaped by a daily struggle to find water.[2] Families that were once the refuge of the poor disintegrate. Plants and animals that once lived in fertile lands suffer too, and many of them become extinct. As weapons of war become more available and more deadly, more and more people live in terror or flee ancestral lands to seek safety. If life is meant to be enjoyed by all individuals in a world supported and sustained by Goddess/God, human beings have a lot of work to do to transform patterns and structures of injustice that we ourselves have set in motion.

For process philosophy, human beings are not the only ones with the capacity for joy. When Hartshorne asked if birds enjoy singing, he was also asking if individuals other than human beings enjoy life. Hartshorne had no quarrel with those who say that birds sing in order to claim territory and attract mates. But he felt birds must enjoy singing; otherwise they would have no need to continue doing it so frequently and so beautifully when territory is secure and mates are found. Nor would some birds develop such intricate songs that they are appropriately viewed as musicians.[3] The ancient Greeks recognized that when they made music they were imitating birds.[4] If birds enjoy singing, then dogs must enjoy running, and dolphins must enjoy jumping out of the sea. Who knows, even atoms may enjoy whirling in space.

After having spent many years studying sea life, ecologist Rachel Carson wrote: "Contemplating the teeming life of the shore, we have an uneasy sense of the communication of some universal truth that lies just beyond our grasp. What is the message signaled by the hordes of diatoms, flashing their microscopic lights in the night sea? What truth is expressed by the legions of barnacles, whitening the rocks with their habitations, each small creature within finding the necessities of its existence in the sweep of the surf? And what is the meaning of so tiny a being as the transparent wisp of protoplasm that is sea lace, existing for some reason inscrutable to us—a reason that demands its presence by the trillion amid the rocks and weeds of the shore? The meaning haunts and ever eludes us, and in its very pursuit we approach the ultimate mystery of Life itself."[5] Carson chose these words for her eulogy. Her death prevented her from writing the book she contemplated about Life itself.[6] What is the mystery that Carson intuited but could not quite put into words? Does the relentless quest for the continuation of life in forms diverse and beautiful have something to do with joy?

From many perspectives the statement that life is meant to be enjoyed is the most radical one in this book. We have been taught that life is difficult. "It is a dog eat dog world," we are told. "You must always be on the lookout for the other dogs." "Time is money," we are told. "Do not waste it." "This world is a vale of tears," we have been told. "Those who suffer are blessed by God." "It is better to give than to receive," we have been told. "Always put others' needs first. Parents must sacrifice themselves for their children," we are told. "Women must sacrifice their own needs to keep the peace at home," we are told. "God will reward you in heaven." From all of these perspectives, the idea the life is meant to be enjoyed is incomprehensible. "Those who enjoy themselves too much," we are told, "are lazy and self-centered. They will never get ahead," we are told. "Those who enjoy themselves too much," we have been told, "have no concern for others."

In sharp contrast to this, a frequently repeated invocation to the Goddess states that "All acts of love and pleasure are My rituals."[7] When I first heard these words my body tingled. "Could it be, I wondered, that the divine power wants us to enjoy life?" This idea was contrary to so much of what I had been taught about religion. I had been taught that religious acts were those in which the other comes before the self. I had been taught that religious acts were about not thinking bad thoughts about others and about helping those less fortunate than myself. I had not been taught that pleasure was evil, though some of my friends had been taught exactly that. Yet I had never been taught that pleasure might be considered holy.

Traditional Christianity has a different view of physical pleasure. Christianity teaches that sin and death entered into the world through the body of a woman. This woman is Eve. Her naked body is the devil's gateway. All who are born of woman will be subject to death. Following Plato, many Christian philosophers and theologians imagined that man's true home is an immortal realm. If Eve had not sinned, man would have lived forever in paradise in the presence of God. For some Christian theologians life in the body is itself the result of sin. Many claimed that the lust (and pleasure) involved in the sexual act transmitted the sin of Eve from one generation to the next. Christ came into the world to liberate us from sin. Through his sacrifice, death is overcome. By following Christ we can become immortal.

This is a very strange picture. Feminist theologians have pointed out that it unfairly places all the blame for sin on the woman. The nakedness of the woman makes it logical to assume that sin has something to do with sexuality. (No wonder we are afraid of enjoying life.) Femaleness, nakedness, and sexuality must be bad. This despite the fact that we are all conceived in, nurtured in, and born of a female body. In this story death is imagined to be a great evil, something that humanity would not have suffered, if the woman had not sinned. Femaleness, nakedness, sexuality, and death are somehow connected. Feminist theologians have questioned the habits of thinking that link femaleness, nakedness, sexuality, and death. What if femaleness, nakedness, and sexuality are good? What if death is not evil, but simply part of life? What would it mean to celebrate and enjoy femaleness, nakedness, and sexuality? What if life is to be enjoyed and death is to be accepted and celebrated as part of the cycle of life? For process philosophy, these are the right questions.

Christian tradition is not alone in associating the female body with sexual pleasure and sexual pleasure with death. Wherever men have equated spiritual perfection with asceticism, women and the female body have been vilified. Yet men as well as women have sexual urges and experience sexual pleasure. Men as well as women die. So why is the blame put on women? Feminists suggest that men have projected their own fears and vulnerability onto women. By making women the scapegoats for finitude, men imagined that they could escape the body and death. To its credit, process philosophy resists this temptation. Process philosophy accepts death as part of life. It also affirms sexual pleasure to be one of the enjoyments inherent in embodied life.

What does it mean to enjoy life? What does it mean to enjoy a life that ends in death? To enjoy life is to participate fully in it. To experience each moment to the fullest. In our bodies. To see, to touch, to taste, to smell, to hear. To delight in the existence of others. To delight in our own life. To appreciate beauty and diversity. To feel joy and sadness. Buddhist meditation teaches us to be present in the moment. Humanistic psychology, influenced by Buddhism, does too. Why is this so hard for so many of us? The reasons are complex. But part of it is this: modern western culture does not truly value life in the body. Parenting styles reflect this. Children and other living beings are not cherished for their individuality, spontaneity, and creativity. Parents value achievement and conformity to patterns of behavior they choose, whether these are traditional or countercultural. This method of parenting is to some extent patterned on traditional religious understandings that life is about following certain rules. Children learn to distrust the messages that come to them through their bodies. This is where the damage begins. To recognize this is not to blame parents. All parents once were children.

Many of us remember being punished when we did something "bad"—for example, not finishing our homework or not coming home on time. What we may not remember is just as true. Many of us were disciplined when what we were doing was enjoying life too much. When we were laughing and giggling, when we were jumping and shouting, when we climbed trees or cannonballed into the swimming pool. There was nothing dangerous or wrong in what we were doing. Perhaps we disturbed the peace our parents wanted for themselves. Maybe our parents were afraid of too much feeling. We internalized parental figures and learned to "discipline" our own emotions. Because we were not valued simply for being ourselves, many of us also came to believe that we would be accepted and loved only if we met certain standards. Many of us imagine that we will be loved only if we are perfect. Since perfection is impossible, we criticize ourselves continually and stop ourselves from enjoying all that we do.

Another impediment to enjoying life is the way we organize time. As we all know, life's greatest joys often happen when we least expect them, sometimes when we are sitting around and doing nothing. In order to see the chaffinch that comes to my balcony railing, I must look up from my work. To do that I must have time. For parents to appreciate spontaneity and creativity in their children, they must have time. To enjoy our gardens and the nearby woods, we must have time. To enjoy the pleasures of the body, whether a leisurely bath, daily yoga, or lovemaking more than once a month, we must have time. And yet we have been told not to "waste" time. In traditional agricultural life, nature dictated rhythms of intense work punctuated by times of rest and celebration. Industrialization changed that. Americans have longer workdays and less vacation time than their European counterparts. In Europe it is common to have four to six weeks of summer vacation in addition to other holidays. Most Americans must make do with two. It can take a week or more to begin to relax from the stresses of modern jobs, and now with cell phones and laptop computers, many take their work along with them. We need to learn that time not devoted to work is not time wasted but time of life and renewal. We need to think about ways of changing the conditions of work and play.

Another way we keep ourselves from enjoying life is by imagining that we should be happy all the time and never feel sadness or pain. Yet sadness and pain are also part of life. So what do we do? All too often we practice denial. We do not allow ourselves to feel pain and vulnerability. It is all about being in the body. When we live in the body we are open to whatever happens in and to the body, including pleasure and joy, sickness and dying. If we understand that illness can be part of life, then we do not have to view it as an interruption of life. When we are sick, we can enjoy a time of enforced rest and the freedom to read or to let our minds wander. We can learn to accept help. We can learn to listen to and nurture our bodies. We can enjoy learning how it feels to be ill, thus broadening our sympathies for others. Our culture teaches us not to grieve. But grieving too is part of life. If we accept that life includes loss, then we can allow ourselves to feel sad. When we grieve fully and passionately, we are also living fully and passionately. We can even enjoy learning what it feels like to know great sadness, again widening our sympathies. Dying is part of living. If we can accept our dying or that of others and not fear it, then we can enjoy whatever time is left as much as possible. Part of that enjoyment will be learning what it feels like to be in the process of dying.

I am not suggesting that feeling is everything or that we need to "let it all hang out all the time." There are appropriate times and places in life for immediacy and others for reflection and anticipation. What I am saying is that in modern cultures the balance has swung too far in favor of the mind controlling the body. What we as a culture need to learn is to listen more to the body and to enjoy the world through our bodies more. Obviously I would not be sitting at my desk writing these words if I were living totally in the moment. I began with a vague plan for this book, shared it with others, started writing, developed an outline. I have already spent countless hours reading, discussing, and reflecting on the ideas that I am writing about at this moment. What you cannot know unless I tell you is that for the most part, I have enjoyed the process of writing this book. There have been times when I made myself go to the desk and work, but most of the time I wanted to. I wrote the first draft of the first half of this book in the mornings and early afternoons of two summer months. During that time I spent most late afternoons at simple seaside taverns and in the sea with a friend. We laughed and swam in the crystal clear or agitated sea and feasted on fresh fish and tomato and cucumber salads washed down with water and wine. Even to write these words is to risk ridicule. "Who does she think she is," someone might say, "swimming in the sea in Greece while I am stuck here working so hard? What possible value could a book written by such a hedonist have?" Anyone who is having a good time is suspect in our culture. Instead of asking why we do not enjoy our lives more and how we can change our situations, we sometimes find it easier to criticize others. This way the status quo is not upset.

Anyone who dares to propose that we might enjoy life more runs the risk of being characterized as a hedonist or a narcissist. Hedonism is understood to mean a life devoted to the pursuit of pleasure and the gratification of the senses, calling to mind the orgies of Roman emperors. Narcissism refers to focusing on the self to the exclusion of others. Those who criticize hedonists and narcissists portray themselves as responsible, hard working, disciplined, and socially concerned. In this setting too, it is difficult to imagine that life is meant to be enjoyed. Either you enjoy life and care nothing for others, or you don't enjoy life too much but you care about others. Is there something wrong with this picture? The hidden assumption is that enjoyment is inevitably selfish and self-serving. Need it be?

What if enjoyment is not individual but relational and social? What if joy is inherently shared and sharable? What if feeling joy is not at the expense of others but with and for others? "If that is so," you might be thinking, "then how are we to understand the Nazi leaders whose enjoyment of classical music did not stop them from killing and torturing others in the death camps?" For process philosophy, the world is relational and social through and through. This means that joy is never an individual experience. Joy is not a feeling "I" have. Joy occurs in relationship with others, both human and non-human. I feel joy when the sun peeks through the clouds after a soothing rain, when a little black dog named Babis jumps at me, when someone smiles. Joy is evoked in relationship. Could some Nazis enjoy the sound of music while ignoring the cries of the inmates of the concentration camps? Apparently they did, or at least they said that they did. What are we to make of this? We might hypothesize that the Nazis did not enjoy music fully in their bodies, but rather viewed it intellectually as an example of the superiority of Germanic culture. But the problem is not that the Nazis enjoyed music. The problem is that they apparently were unable to feel the feelings of other human beings. Or perhaps did not act on the feelings they felt. The reasons for this failure are complex and include the fear incited by the Nazi regime, methods of discipline in early twentieth-century German families and schools, military training, long-standing currents of Christian anti-Semitism, defeat in World War I, economic deprivation, and above all the influence of Hitler. We need also to remember that what happened in Germany in the Nazi era can happen to any people or group if circumstances and choices coincide. The Nazis have shown us that on a certain level at least, the enjoyment of some aspects of life is compatible with great cruelty. But enjoying life in the body in the fullest sense is not compatible with lack of concern and compassion for others. If we truly enjoy life, we enjoy all our relationships and sympathize with the experiences of a wide variety of individuals, both human and nonhuman. When we feel the feelings of others, we will want to change the personal, interpersonal, and structural factors that cause them to suffer.

Another important factor in the enjoyment of life is interest. Reflecting on her long marriage to Charles Hartshorne, Dorothy Hartshorne asked a young woman, "What is it that you must never do in a marriage?" The young woman thought for a moment and replied, "Never hurt each other?" "Oh, no," Dorothy replied, "You will hurt each other all the time. What you must never do is bore each other!"[8] Charles Hartshorne made similar comments in his works.[9] I do not know if Dorothy influenced Charles or Charles influenced Dorothy, or if they arrived at this conclusion together. In any case, it is an important insight. Those who enjoy life are interested in it. Sometimes we think of relationships as two becoming one or in terms of different people sharing a fundamental oneness. Then we find difference threatening and downplay it. But in a changing world, difference creates interest and interest can lead to enjoyment. Those who are fully alive appreciate diversity and difference and are ever eager to learn something new about someone or something. Hartshorne was aware that too much difference is chaos. There must be an element of continuity in experience for it to be comprehensible. However, when we recognize that difference is what makes life interesting, we can choose to appreciate difference rather than trying to smooth it over or root it out.

We have been taught that talking about others is gossip. Gossiping is something women do too much of and it is bad. Of course some talk about others with the intent to harm. But much of what we call gossip is a way of expressing interest in the lives of others, of sharing and sympathizing with lives other than our own. Gossiping is one way of enjoying life. One of my friends has jokingly yet tellingly suggested that we need a "thealogy of gossip."[10] Another way of putting this is to say that expressing interest in the lives of others can be a sacred activity. Talking about others often helps us to understand ourselves better and can widen the range of our choices.

I have long been interested in the similarities and differences in the ways the Greeks of my traditional village look at life as compared to my own suburban and urban American way of looking at things. Recognizing the different ways my Greek friends approach the physical realities of life, I have learned new ways of living in my body. In recent years two new areas of interest have opened up for me. One is birdwatching. I now can identify several hundred species of local birds and am beginning to understand when and where they are likely to be seen on my island. The other is geology. I enjoy thinking about what it means to live on a volcanic island. I want to learn more about the multicolored stones on the beaches, the rosy volcanic rocks used in traditional building, and the rugged boulders that define the landscape.

These interests have enabled me to appreciate the complex ways in which my body is part of the land. They have sparked renewed commitment to environmental activism. The world is full of so many interesting things that there is no reason to be bored. Yet boredom is widespread in modern life. Process philosophy suggests that the reason for this is that we are not fully present in our bodies and in relationships with each other and the world. With modern life becoming increasingly violent and ugly, it requires courage to stay open to the world.

Those who say that the meaning of life is enjoyment are often said to have an "aesthetic" rather than a "moral" approach of life. An aesthetic standpoint calls attention to beauty and enjoyment, while the moral standpoint to which it is contrasted focuses on following the rules, often out of a sense of duty. Whitehead and Hartshorne characterized their philosophies as more aesthetic than moral in order to emphasize that they did not accept the common view that the meaning and purpose of life is a matter of following rules or acting out of a sense of duty. However, the distinction between aesthetic and moral should not be taken to mean that the process view is less ethical in the deepest sense than those it criticizes.

An aesthetic view of the world is one that takes its cue from our experiences of the appreciation of beauty. The word "aesthetic" stems from a Greek word meaning "feeling." "An-esthesia" is given to stop feeling. The aesthetic attitude to the world is the ability to feel deeply about life, to be moved by it. Appreciating the colors of a sunset, the grace of a child climbing a tree, or the warmth of a lover's smile are aesthetic experiences. Aesthetic experiences can come through listening to a symphony, watching a dance performance or a play, or looking at a painting. For those with the ability to feel deeply, such experiences reveal life's most profound significance. When I was a child floating in the sea and looking up at the sky, I felt at one with the universe and at peace with the world. In my mid-twenties, watching "Dances at a Gathering" by Jerome Robbins set to the piano concertos of Chopin moved me deeply. Men and women dressed in different colors explored a wide variety of relationships through dance—male and female, male and male, female and female, groups, solos. This dance expressed feelings I had about the great beauty of relationships. The common denominator in these experiences is a deep feeling of enjoyment of the beauty in life.

Although we often think of aesthetic experiences as the appreciation of beauty, Whitehead and Hartshorne understood the aesthetic experience to include appreciation of all the complexities of life, including tragedy. At nineteen I spent many hours looking at Rembrandt's self-portraits. I was not responding to beauty in a superficial sense, but to the deeper beauty that is created in the expression of the complexity of human suffering and survival. Similarly, while celebrating youth, joy, and eros, "Dances at a Gathering" did not exclude experiences of separation and longing for inclusion. Whitehead and Hartshorne believed that enjoyment of life that includes suffering can be understood on the model of aesthetic appreciation. We can appreciate painful experiences in our lives, though we would not necessarily have chosen them. My mother's dying was painful for me, and yet it changed my life. We enjoy tragedies as much as comedies because both help us to understand our experiences. Process philosophy says that a certain amount of suffering creates contrast in life, enabling us to enjoy it more. Yet it is important not to take this line of thinking too far, or we will end up imagining that all pain and suffering are part of the divine purpose. This is not the process view, as we have seen in the previous chapter.

The aesthetic standpoint has sometimes been understood to privilege so-called high art over popular and folk culture and the creations of the human mind over the creations of nature. Thus listening to an opera might be understood to be more important than delighting in birdsong. Or visiting an art gallery might be said to be more valuable than hiking in the mountains. Or looking at a classical Greek temple might be considered more significant than studying the way a farmer fit stones together without mortar to build a terrace wall. If this is so, then we would have to say that those who have not been educated to appreciate the traditions of so-called high art are missing out on an important dimension of life. But this cannot be true. Appreciating beauty and complexity within nature and in the lives of others is the primary form of aesthetic appreciation. Creating and appreciating art is modeled on appreciating life. As my Cretan friend Mr. Nikos who loved the Skoteino cave was fond of saying, "Human beings have never made anything to compare with what nature makes."

An aesthetic approach to life is in the first instance interested and appreciative rather than judgmental. It is open to diversity and difference. Indeed, it finds diversity and difference interesting. When confronted with something new or different, a person taking an aesthetic approach to life will not ask, "Is this good or bad?" But rather, "What is there in this to appreciate?" A person with an aesthetic view of life is likely to be more tolerant and more open to being influenced by another experience, another person, another culture, or another landscape than one who first looks at the world from a moral perspective. The moralist is more likely to imagine that he or she knows how things should be and to judge others for not conforming to certain standards of behavior or ways of doing things. For the moralist life is a series of either-or choices: "Either the natives should be wearing clothes, or we all should be going naked." "Either we do it our way, or all hell breaks loose."

The moral view to which the aesthetic view is contrasted is a particular sort of moral view, one with which many of us are familiar. This is the idea that doing well has to do with following the rules. In this paradigm, "you do as you are told," whether by parents, teachers, leaders, military commanders, God, or the principles of universal reason. In this sort of a moral view, doing the right thing is often portrayed as "ignoring your feelings." "You may not want to do it," most of us have been warned, "but you are going to do it anyway, and you are going to do it now! Why? Because I said so! Don't defy me." According to this view, the unruly feelings of children and animals must be educated, disciplined, or beaten out of them. Similarly, adults must keep close control over their feelings in order to live in society with others. This is the view that Whitehead and Hartshorne rejected when they called their philosophies aesthetic rather than moral. Whitehead and Hartshorne believed that ethical judgment must be rooted in appreciation and interest in living. As Hartshorne said: "It is sad to think of parents who worry about teaching their children morals . . . yet have scarcely a notion of the prior necessity to help them find life interesting and enjoyable. Desperately bored or unhappy children are not good candidates for moral instruction, especially by those who bore or painfully annoy them."[11]

Many of us reject the moral view when it is presented to us in terms of the discipline of parents and teachers. Yet this view is deeply rooted in our culture and in us. The difference between a moral and an aesthetic view of the world in the deepest sense is about how we feel about feelings and enjoyment of life. Are feelings and enjoyment to be trusted or distrusted? Is feeling all of our feelings a good thing? Or are there feelings we must control or suppress in order to get on with the business of life or, more importantly, to do the right thing? What if I want to make love with my neighbor's spouse? What if I want to steal from my employer? What if I want to lie on the beach all day, every day? What if I want to beat my spouse? What if I enjoy being beaten by my spouse? What if the rich enjoy oppressing the poor? Obviously feelings of enjoyment are not an infallible guide to what is right and wrong for us and for others. Our feelings have been shaped by experiences in our families, societies, and cultures. Because feelings can be unreliable, moralists often urge us not to trust our feelings at all. "Trusting our feelings," they say, "will only get us into trouble." But, Hartshorne, Whitehead, and I would counter, doing what we are told has probably gotten people into more trouble than trusting our feelings. All authoritarian systems work because people have been taught to follow orders, even when their feelings tell them to do otherwise. Nazi Germany is only the most extreme example of widespread patterns of behavior that exist in every culture in which habits and structures of domination exist.

Feelings are not innocent, because they have been formed by relationships, societies, and cultures. This is fully consistent with and indeed anticipated by the process view of the self as formed in social and personal relationships. In advocating an aesthetic approach to life, Whitehead and Hartshorne are not saying that feeling or enjoyment is the only thing to be considered in ethical decision-making. What they are saying is that unless we are interested in and enjoy life in our own bodies, it is unlikely that we will be able to be taught to appreciate and enjoy the lives of others. And if we cannot appreciate and enjoy the lives of others, it makes no sense to imagine that we will exercise care and concern in relation to them. Interest, joy, and sympathy are deeply related to each other, and they in turn are the basis of genuinely ethical behavior.

In the process view of the world, the divine power is not a lawgiver. Obedience to divine will, word, or law is not what Goddess/God is seeking from us. This is a major departure from what many of us have been taught about religion. What then does Goddess/God want of us? Goddess/God wants us to enjoy life for however short or long time we live. Goddess/God wants our enjoyment of life and sympathy with others to motivate us to seek to create a world in which other human beings and all beings in the web of life can enjoy life as much as possible. The more we enjoy life, the more joy there is to share with others including Goddess/God.

According to process philosophy, the purpose of all life in the universe, including divine life, is enjoyment. Process philosophy differs greatly from many other religious philosophies in focusing on enjoying this life rather than on seeking life after death. Hartshorne named the doctrine of immortality one of the common theological mistakes of classical theism. Yet for many people, religion is synonymous with belief in life after death. For some Christians, the promise of reward in heaven and the fear of punishment in hell are among the primary reasons for believing in God. For some followers of non-western religions, the promise or threat of karmic rebirth in a higher or lower body is an important aspect of religious life. Saints in both the east and the west have sought suffering in the hope that it would purify the soul. Non-believers claim that religions and religious authorities gain power over people and turn their attention away from life in this world by activating hopes and fears about things that could not possibly be true. Most atheists agree with many believers that religion is about immortality. Belief in immortality is tied to the first theological mistake, the notion that God exists in an unchanging realm which is more perfect than the world that we know. Religions that focus on immortality ask us to give up the fleeting pleasures of a changing world and the body in exchange for immortal and changeless bliss.

In contrast, process philosophy values the changing world. Our goal is not to escape the world but to participate fully in it. From a process point of view, there is only one being in the universe who is absolute, infinite, and immortal, and that being is Goddess/God. To be born, to live for a time, and to die is an inherent part of the lives of all individuals other than Goddess/God. Yet Goddess/God does not live in an immortal realm apart from the changing world. The changing world is the body of Goddess/God and therefore is sacred. The purpose of divine life too is to enjoy the changing world.

Why does process philosophy say that immortality is not the goal of human life? For one thing, process philosophy disagrees with traditional understandings of immortality that connect it with divine judgment as an aspect of the moral view of life. "Immortal bliss in heaven," it has been said, "is the reward for following the divine will while eternal suffering in hell is the punishment for not following the divine will." For process philosophy (as for Saint Teresa of Avila and others), the notion that the divine power would condemn individuals to eternal suffering makes a mockery of divine sympathy and love. While human sympathy and love may never be unconditional, divine sympathy and love are. Goddess/God would never condemn anyone to eternal suffering. This view has been deemed the error of "universalism" by some Christian theologians. This (alleged) heresy is that of believing that God would not abandon anyone. Universalists believe in universal salvation.[12] Process philosophy agrees with their logic. As long as any individual is suffering or causing others to suffer, Goddess/God is with that individual transforming his or her suffering through understanding it and attempting to persuade the individual not to cause unnecessary suffering to others. This is what divine sympathy and love mean. They simply are not compatible with the doctrine of eternal punishment in hell.

In addition, for process philosophy, the notion that human beings can be motivated to do good rather than evil only by the promise of heaven or the fear of hell makes a mockery of ethics. It limits human freedom and ethical decision-making to following the rules and tends to produce rigid and unhappy people. If we truly understand that the consequence of any particular choice is reward in heaven or punishment in hell, then our freedom is analogous to that of a rat in a maze that has figured out that going left means food while going right means a painful electric shock. The rat is still free to choose to go right, but the freedom of the rat has been severely limited by putting it in a maze in which the consequences of its actions are predetermined. The life of a rat in the wild is far more complex and interesting.

Our lives are more complex than moral theories emphasizing reward and punishment can account for. Yet moral systems based on reward and punishment deny the complexity of life by asserting that moral behavior is about following a number of rules. "You should never have sex outside of (heterosexual) marriage." This is a rule that in principle can be followed, but it may have little to do with living an ethical life. "You should always put the needs of others before your own." This rule, if followed perfectly, would probably mean starving to death. In the lives of many it has meant denying or repressing legitimate needs for care and concern in this life. There are many people who believe that they are good but who do not enjoy life and cannot teach others to enjoy it. For process philosophy, a reward-and-punishment-based morality starts off on the wrong foot. Truly ethical behavior is motivated by sympathy, not by fear of punishment or calculation of reward.

More importantly, for process thinking, body and soul or body and mind are understood to be one continuum. We know that the body dies. There is no reason to expect that the mind or the soul can live apart from the body. Experience and modern science tell us that death is the ending of life in the body. Process philosophy sees no reason to disagree. Humanistic psychology and feminism agree with process philosophy and modern science that our bodies are ourselves. From this perspective it does not make sense to assert that actually, our souls are ourselves and our souls are immortal.

What about life-after-life experiences? The tunnel and the white light? Shamanic journeys out of the body? Communications in dreams and visions from people who are dead? Past life experiences? Don't these suggest that the soul is independent of the body, thereby in some sense justifying traditional religions that focus on life after death rather than enjoying this life? My eighth-grade teacher told our class that his mother died or almost died while giving birth to him. She had the sensation of floating above her body and being called back to it by her child's cry. Such experiences have been called "life-after-life experiences"[13] and are often said to involve entering a tunnel and then coming into a white light in which love, loved ones, or spiritual guides are seen or felt to be present. Shamans say that their spirits are able to leave their bodies and travel to other places while their bodies remain in a deathlike state. Many people have reported seeing loved ones in visions or intense dreams at the time of their deaths. Often the one who reports such an experience insists that she or he did not know that the person who appeared in the dream or vision had died. Others receive communications that seem to be from the dead years later. More than a year after my mother died, I had an intense dream in which she spoke to me as if from the other side. Some of my friends have had what are called past life experiences. Like many others, they were put into trance and had vivid impressions of having lived in another time or place. One friend felt that she had lived as an Egyptian priestess, while another thought that she must have been a sacrificial victim in an Aztec ritual. What are we to think about this?

As far as I know, Hartshorne and Whitehead did not systematically explore any of these phenomena. The one certain thing that can be said is that all of these experiences are remembered and reported by individuals who are embodied. No one who is still living in a body can say for certain what will happen after the body dies. A degree of uncertainty, agnosticism, and openness to the unknown is appropriate. Still, I am inclined to believe that the experiences described above tell us more about the complex nature of life in the body than about life after death. The life-after-life experiences of the tunnel, the white light, and love have been interpreted as evidence that the soul survives the body. In fact, they may be telling us only about the moment just before death. What happens after that remains unknown. The journey toward the light may conclude in the merging of individual energy with the light, or, in process terms, with divine love that is the ground of all being and becoming. Life-after-life experiences may tell us about the ending of individual existence rather than about beginning another form of individual life. Shamanic experiences suggest that it is possible for the mind or soul to separate briefly or partially from the body. Such experiences do not tell us that mind or soul can survive indefinitely and certainly not forever apart from the body. Visions of or communications from the dead or the dying have been given psychological explanations. It may be that the desire for communication with a person who is dying or has died creates the feeling that she or he has visited us or spoken to us in dream or vision. I am more willing than Hartshorne apparently was to wonder if there are ways that individual spirits or souls can survive death of the body for a period of time through connection to the bodies of those who remember them. If this is so, the dead may be able to communicate with those they have loved. Still, like Hartshorne, I see no reason to hope or believe that any individual soul might live forever apart from its body.[14]

While western religions have imagined immortality in a heavenly realm, eastern religions more often speak of karmic reincarnation. Here the claim is made that "the same" individual can be said to exist in a plurality of different bodies. Theories of karmic reincarnation are often used to explain suffering and elicit moral behavior. "If you are suffering in this life," it is said, "you are being punished for acts committed in a previous life. If you are good in this life, you will be rewarded in the next." Insofar as theories of karmic reincarnation are tied to ideas about reward and punishment, they too may simplify moral complexity and constrain human freedom. Like religons that promise immortality, those that focus on reincarnation may keep us from fully enjoying and appreciating finite embodied life. Moreover, for process philiosophy, theories of reincarnation fail to understand and accept the limitations of life in the body. Process philosophy, as we have seen, denies the strict identity even of individuals who inhabit the same body. I am and am not the child who climbed my grandmother's peach tree, and so on. If we think of the self as a body-mind continuum, it is hard to imagine how "my" self could exist in a body and with a history utterly different that the one I have. Unless there is a part of me that is absolutely unchanging, it does not make sense to say that "I once was an Egyptian priestess" or that "I might come back in another life as a dog." Hartshorne commented wryly on such beliefs. "Those who want to go on being themselves forever and yet pass on to additional experiences after death are either asking for unbearable monotony, endless reiteration of the same personality traits, or they are asking for a unique prerogative of God, ability to achieve self-identity through no matter how great and diverse the changes and novelties. Unconsciously they either want to be bored to death, so to speak, or to be God."[15] For Hartshorne being bored to death is a fate worse than actual death. I agree with him that it is appropriate to think of all lives (except that of Goddess/God) as having beginnings and endings, which is to say as finite. Our finite lives are immortal in the sense that they become part of the life of Goddess/God and are remembered forever. If individual identity does not continue in other bodies, then what have been called past life memories would be understood symbolically rather than literally, and the information they offer would be about this life, not another one. Alternatively they might be interpreted as tapping into some kind of collective unconscious or memory that is not the memory of the person who reports having lived in another body, but the memory of someone else.

If our lives are finite and will end in death, does this mean that everything we do is ultimately meaningless? Many thinkers, both religious and secular, have taken this view.[16] This conclusion is the logical outcome of the overly individualistic tendencies of western thought. This becomes clear when we ask the question: Meaningless for whom? The answer usually given is: Meaningless for the individual who at some point will no longer exist. If we ask the question of meaning from a social and relational point of view, we will reach a different conclusion. A mother's love will have great influence on her child. A teacher's creativity may challenge thousands of students. A neighbor's sympathetic ear may change the life of a lonely teenager. The acts of Gandhi, Martin Luther King, Jr., Rachel Carson, and Rosa Parks affect our lives today. Influence is both positive and negative. Cruelty as well as love is passed down through the generations. A mother's inattention, a teacher's racism, and a neighbor's drunken rage also influence the lives of others. The effects of the pesticides we use in our gardens will outlive us. The meaning of our lives is the web of influence we have created. Yet at some point most of us will cease to be remembered by the living. Does this mean that our influence stops? Not really. The choices we have made will continue to be felt by others. I do not remember my ancestors who left France, Ireland, Scotland, Belgium, Sweden, and Germany, and eventually came to America.

But their decisions to do so have affected my life and the lives of countless others, including the lives of Native Americans who are living today. In addition, our bodies return to the earth, becoming compost for other lives. If we are cremated, our ashes may be returned to the sea or a place we loved. If we understand life as change and relationship, this could be a comforting thought. I will not live again, but parts of my body will form the building blocks of other lives.

If there is no personal immortality or reincarnation, does this mean we should fear death? From a process point of view, the answer is no. There is nothing in life or death that can separate us from the love of Goddess/God.[17] Goddess/God will be with us at the moment of death as Goddess/God is with us at every moment of life. The sense I had when my mother died that she was going to love is perfectly compatible with process understanding. When the individual life ends, it will become part of the divine love that is the ground of all being and becoming. The memory of our individual lives will be preserved by Goddess/God.

Process philosophy distinguishes between what it calls "objective" and "subjective" immortality. "Subjective" immortality means that I as the "subject" of my life will go on living in another state or in another body. The reasons process philosophy does not consider this likely have been discussed above. "Objective" immortality means that my individual life will not be forgotten. In the inner life of Goddess/God, our joys and our sufferings are taken up and transformed in a sympathetic understanding far greater than we could have for ourselves or for anyone else. In the memory of Goddess/God, every single individual in the universe will live forever. Objective immortality in this sense was crucial for Whitehead and Hartshorne, who felt that unless remembered, our lives could be deemed meaningless. For me, it is more important to know that in the body of Goddess/God the elements and energies in our bodies will continue to live in other bodies as long as the earth or the universe survives. I may not survive the death of my body for long, but the elements and energies of my life will continue to be taken up and transformed in other lives in the body of Goddess/God. From this perspective, our responsibilities are not only to ourselves and to the living, but also to those who will come after us, as well as to Goddess/God. Everything we do and do not do will have an impact on the world, which is Goddess/God's body. Everything we do and do not do will also have an impact on the inner life or feelings of Goddess/God.

A friend of mine said recently that perhaps the best way to think about death is not to think about it at all because thinking about it makes life seem unbearably sad. Process philosophy suggests that a better course is to accept and acknowledge that death is the ending of every finite life. To enjoy life, knowing that it will end. To be prepared for death to come into our lives at any moment. Otherwise, when confronted with death we will feel angry and confused because something unexpected has happened. Our suffering is greater when we imagine that death should not have happened to us and to our friends and relatives. We will all die. We can ease the suffering of the dying with care and concern as well as with traditional remedies and modern drugs. We can also become interested in and learn to enjoy the process of dying as much as possible.

When my aunt Mary Helen was diagnosed with cancer, she elected not to have the chemotherapy and radiation treatments that my mother had suffered two years previously. My mother, who was not one to question authority, privately referred to these treatments, including the excess dose of radiation that burned her, as "torture." Having learned from her sister's experience, Mary Helen accepted that she was dying and spoke openly about her imminent death with family and friends. At the same time, she filled her last months of life with joy. I remember her planting bulbs in pots in January and wondering out loud if she would be around to see them bloom. If we accept that death is the appropriate ending of life, we can enjoy the time we have. And we may not find it necessary to prolong it with painful medical treatments or with tubes and machines.

Process philosophy does not deny suffering. Indeed, because of its focus on this life and divine sympathy, process philosophy recognizes the tragic character of life more clearly than many other world-views do. Process philosophy places the responsibility for alleviating humanly created suffering in human hands, as I will discuss in chapter seven. For process philosophy, suffering is felt and shared and remembered forever with great sadness by Goddess/God. There is no other life for us in which our suffering will be redeemed. Nor is there any promise or guarantee that our individual lives will be long or short. Therefore we should make the best of the time we have and do what we can to increase the joy of others.

For process philosophy, changing life in changing bodies is meant to be enjoyed. Joy is evoked in relationship and is meant to be shared. But if enjoyment of life in the body and in the changing world is the purpose of human and divine life, then why have so many religious philosophies not recognized this? Why have so many religions and philosophies focused on immortality and reincarnation? Why have we been told that we can or must escape this life? It is hard to resist wondering if the source of the problem is a rejection of embodied life that begins with rejection of the female body, the body through which we came into this world.

Chapter Six

Embodied, Embedded Knowing

A question that arises for all of us when we think about the meaning of life is: How do we know? Do we trust the body? The mind? The spirit? Great thinkers? Favorite authors? Friends? Communities? Revealed traditions? And if so, which one or which ones? How do we know what is true? For process philosophy, all knowing comes to us through the body, which is embedded in relationships, both personal and social. Our perceptions are shaped and limited by our bodies and our personal and social locations. If we are open to learning from other individuals, from the natural world, from our bodies, and from Goddess/God, our perspectives become wider. If we are not, we remain narrow, our perspectives shaped and controlled by conventional thinking. Yet no matter how open and interested we are, our knowing will always be fragmentary. Our bodies and the world body are changing from moment to moment. Thus our understanding too may change, develop, or deepen. No one of us will ever have the whole picture. No one of us will ever know the truth, the whole truth, and nothing but the truth. All of our knowing is in some respects uncertain. From one perspective, uncertainty makes life interesting. There are always new people to meet. There is always more to learn. There are always other ways of looking at things. Because we do not know everything, we remain open to others, to life. Changing life asks us to be flexible and open in our thinking.

Yet uncertainty can be unsettling. Some of us are afraid of asking too many questions and cling to what we have been taught. Others ask questions but fear the consequences. "What if we base our whole lives on the wrong belief," we say, "or devote our whole lives to the wrong cause?" In order to solve the problem of uncertainty, many traditional religious thinkers have appealed to authority. "Because we cannot know," they say, "we must have faith." "Because we do not know what we should do," they say, "we ought to submit our wills to higher authority." Various authorities are proposed, including books, persons, and traditions. Because such authorities are said to be inspired by God or to live in direct connection with divine spirit, we are asked to put our complete faith and trust in them. In some cases threats are made. Those who do not follow traditional authorities may face shunning, excommunication, stoning, or the threat of eternal punishment in hell. Those who do not submit to traditional authorities will not be saved or will be denied the teaching that leads to enlightenment. Those who claim to have infallible revelation often compel others to agree with their ideas, for they cannot bear the uncertainty created by knowing that others have a different way of looking at the world. For process philosophy, the notion that we should submit our wills to authorities in exchange for intellectual, moral, or spiritual certainty is a mistake. There simply is no way of getting around the relativity and fragmentariness of knowledge. Hartshorne calls belief in infallible revelation one of the six theological mistakes. Though his use of the word "infallible" may suggest that he is talking only about the most extreme forms of spiritual authoritarianism, in fact he is speaking about the broader problems of a quest for certainty or desire for absolutes in a changing world.

Why does process philosophy insist that it is a mistake to submit our will or mind or body to any authority? There are two kinds of answers to this question. One is pragmatic. The other is theoretical. The pragmatic objection is that when we look at traditions that have claimed infallible revelation, the record does not look very good. The theoretical objection is that it does not make sense to say that what is finite can be anything other than limited, which is to say imperfect or fallible to some degree. As we will see below, all claims of infallible revelation in texts, persons, or traditions come up against problems regarding authority and interpretation. Far from solving problems, the claim to infallible revelation creates them.

The historic record of traditions alleged to be infallible is not necessarily praiseworthy. Most readers of this book probably think it is wrong that some traditions within Islam teach that those who die in terrorist acts in defense of the faith will be immediately taken to paradise where they will feast on watermelons and grapes and rest in the arms of virgins. Yet in Christian tradition, those who fought in the Crusades were promised much the same thing. Many North Americans and Europeans find it appalling that some traditions within Islam require women to wear veils. Yet it is conveniently forgotten that Orthodox Jewish women are still required to cover their heads in public; or that until Vatican II women could not enter a Roman Catholic church unless wearing a hat or a scarf; or that Orthodox Christian women are still forbidden to touch an icon while menstruating. Contemporary Roman Catholic and conservative Protestant traditions may not require the veiling of women, yet they are attempting to prohibit knowledge of birth control and access to abortion around the world. In the long run this may do more harm to the cause of women's rights than veiling. During the Inquisition, heretics, women, Jews, and homosexuals were tortured and burned at the stake in the name of faith. At the same time pious Christians were flagellating themselves until their bodies were bloodied in order to purge sin. Though some progress has been made, Protestant and

Roman Catholic authorities still have not transformed the Christian roots of anti-Judaism in the Bible and in liturgy, nor have all Jewish and Christian groups embraced the full humanity of women, gays, lesbians, and transgendered individuals. In the Hebrew Bible, slavery was condoned, as was the secondary status of wives and daughters. In the United States prior to the Civil War, it was assumed by many Christians and some Jews that slavery and the subordination of women were ordained by God. The record of major religious traditions with regard to slavery and the equality of women alone should be enough to convince us that no religious tradition should be considered infallible.

It is often said in defense of allegedly infallible traditions that the "excesses" of certain groups of the faithful must be understood as deviations from the "true" meaning of the faith. Thus, for some liberal Muslims, those who teach that dying while committing terrorist acts will earn a place in paradise have not understood Islam correctly. Or for liberal Christians, the opinions of the Roman Catholic hierarchy about abortion and birth control should not be understood to be the true Christian position on these matters. Some liberal defenders of traditional religious faiths criticize views touted as infallible with which they disagree yet do not question the doctrine of (infallible) revelation per se. Inevitably they appeal to what they call a "true" as opposed to a "false" understanding of revealed truth. Others recognize that the notion of infallibility must be challenged as well.

There are theoretical problems inherent in traditional notions of infallible revelation. Christians, Jews, and Muslims locate authority in the written word of the Bible, the Torah, or the Koran. Traditionally it was asserted that God inspired the writing of these books, speaking directly into the ear of human authors such as Moses or David, the four evangelists, or Mohammed, or guiding their hands as they wrote. Faced with inaccuracies or inconsistencies in the various texts, some will modify this claim. Perhaps God did not inspire the actual words of the holy text or verify every factual reference, but God did inspire the deep, essential, or overall meaning of the text. Or, God inspired Moses or David or Mohammed, but a fallible human instrument transmitted the inspiration. Or, God inspired the authors as they wrote religious texts, but not by dictating every word. If sacred texts are written by fallible human beings living within particular language and cultural situations, then they must contain elements of revealed truth combined with other elements that are culturally relative. For example, Jewish feminists might say that God intended to make a covenant with all of the Jewish people, including women. Moses or other writers of some portions of the Bible followed the conventions of their time in assuming that only men were able to undertake the covenantal responsibility. The notion that God entered into a covenant with the Jewish people is viewed as revealed truth, while the notion of women's secondary status expressed in the same biblical texts is discarded.

Yet there remains the troubled problem of interpretation, as this example shows. If there is a deep meaning in a text, then there must also be a superficial one. If there is an essential meaning, then there must also be an inessential one. If there is overall truth, then there must be errors in detail. How do we know which is which? In other words, how do we know that the meaning anyone finds in a text is the one intended by its author or by God? In order to assure that one interpretation is right while others are wrong, authorities outside the text itself must be appealed to. In many cases the appeal is to an inspired person or persons—priests, rabbis, ministers, imams, or gurus. But individuals may fail, or their perspectives may be limited. What do we do then? Sometimes a religious tradition as a whole is invoked to guarantee the truth of the text. We then have the Bible as understood by the holy Catholic Church, the Torah as understood in rabbinical tradition, and so on. And yet traditions are complex and varied. How do we know which aspects of tradition to follow? How do we resolve contradictions within traditions? Roman Catholic tradition stands alone in asserting without qualification that a single man, the pope, acting in the name of the Church or "ex cathedra" has access to infallible revelation. Whether the powers of other religious leaders (such as, for example, the Dalai Lama) should be understood on the Roman Catholic model is open to question. Wherever a single individual is endowed with infallibility, the question of final authority is solved. With the pope insisting that women may not use birth control and can never become priests and attempting to downplay the pederasty of male priests, one wonders how long the mystique of papal infallibility will hold up. Yet without a single final authority, there is an endless regress. Either many and even conflicting interpretations are allowed, or appeal is made to some authority to unify interpretations or to resolve conflicts.

Sometimes a person rather than a text is viewed as the primary location of religious authority. The identification of revelation with the written word has sometimes been said to be a peculiarly Protestant and to a lesser extent Jewish and Muslim obsession. "A text," it might be countered, "is but words, until its meaning is revealed in the life of an inspired individual." If text or tradition is communicated through a living inspired individual, we can rely on him or her to tell us which parts of tradition are essential and which inessential. More importantly, inspired individuals are not limited to words when communicating spiritual truth. In the Hasidic tradition in Judaism, the rebbe became the focus of attention. Devotees said that they could learn as much from the way the rebbe put on his shoes as from the way he prayed. In some Hindu traditions, gurus play a similar role. Followers of gurus often say that the guru radiates love or peace, and that just being in his or her presence is inspiring or healing. Followers of inspired individuals feel that spiritual truth is often communicated and intuited in nonrational or other than rational ways. For example, spiritual qualities such as devotion, compassion, or love may be more easily sensed in a person than explained in words. Thus for some followers of gurus, text and tradition are less important than the physical presence of the enlightened individual. In Roman Catholicism and in Islam, saints have played a similar role. In

Protestantism, charismatic preachers have been said to open the hearts of others to God. What is the source of such personal religious authority? Usually it is said that the life of such an individual is inspired by God or infused with divine love or that this person has a special kind of relationship with God or openness to the spirit. In some traditions, the role of a guru, teacher, preacher, or saint is a limited one. The individual may be asked to submit or devote herself or himself to an inspired person for a period of time, but the goal of the teaching or training is for the individual to become her or his own spiritual authority. In other traditions, the authority or role of the inspired individual over followers is understood to be long-term or permanent.

Yet individuals who seem to be inspired often have limited perspectives and may fail some or all of their devotees. There is always the temptation to amass wealth or followers or to take advantage of trust. In my opinion the failures of spiritual teachers are sufficient reason to cause us to question whether we should ever submit our wills to spiritual teachers. Protestant preacher Jim Jones led his devotees to mass suicide or murder in the name of God. A well-known writer deeply devoted to the female Hindu Guru Ma was shattered when she told him that his homosexuality was not acceptable.[1] Many individuals, female and male, have been seduced or molested by gurus or priests who claimed to be celibate or by ministers who were married. The violation of trust and the suffering created by being chosen, then abandoned by a supposedly perfect individual can be compared to the suffering caused by incest. If we give up our will to that of a spiritual authority, we are no longer in a position to accept or reject sexual advances freely. When devotees of inspired persons become aware of the misuses of power, many of them blame the victim. Others recognize abuses of power but consider them trivial in relation to the good that the inspired persons are alleged to be doing. Or the abuse of power is attributed to individual failings, not to the mystification of unequal power relations. In this way the notion that there are infallible persons remains unquestioned.

Another source of alleged authority is a particular tradition as a whole. "Religion is not a belief system," it might be said, "but a way of life." "Hinduism is not the guru alone, nor the Vedic scripture alone, but a living practice filled with a tens of thousands of multi-faceted images and tens of thousands of rituals." "Buddhism," it might be argued, "is not about teacher or text, but about the dharma, a way of life rooted in meditative practice and leading to ethical action." "Not Torah and Talmud alone, but Jews united in the daily acts of following the law are the essence of Judaism." "Catholicism is not the Bible, nor even the hierarchy, but a people united by the liturgy of the mass in one holy Catholic Church." In the Goddess movement, authority may be located in pagan or Wiccan traditions[2] or in all the myriad images of the Goddess in western and non-western traditions. Yet here too the questions are: Which Hinduism? Which Judaism? Whose version of Roman Catholicism? Which Goddesses? If appeal is being made to traditions as a whole, how do we know which part of it to focus on? Most of us can find things to admire within every tradition as well as other things that we find strange or even abhorrent. Yet if the traditions as a whole are alleged to be the source of authority, how are we to judge?

If there are so many problems associated with infallible revelation, then why do we keep looking for it? Why have so many of us been willing to give up our power to others? To wish for certainty about things that are uncertain may be a human tendency. To learn to be flexible and open to changing life is a human capacity. But in cultures that define power as power over, as domination and control, learning to be flexible and open is difficult because we have been taught from the cradle that respecting authority is the highest value. Those of us raised to find domination and submission natural will not find it easy to accept uncertainty in ourselves and others. We have been told that someone must always be right. We are afraid of making mistakes. We have not been taught to trust our bodies and the messages that come to us from the natural world. We may not know how to be open and flexible in our thinking and in our relationships with others. Until and unless we learn a whole new way of being in the world, we will always be tempted to turn our power over to authorities that promise to give us the solutions to our problems.

For process philosophy, there are insurmountable pragmatic and theoretical problems with claims about infallible or authoritative revelation. Does this mean there are no spiritual authorities? Are we on our own? In a relational world we are never alone. Power over is not the only power. As we have seen in a previous chapter, there is also power with. There are a few who can learn to play the cello on their own, but most learn from a teacher. There are people with special sensitivity to or training in spiritual matters just as there are people with an eye for color or an ear for music. Such people may express their insights in writing, in spoken words, or in their very being. The words or teachings or paths of such people are gathered together in spiritual traditions. We can turn to them for spiritual insight or guidance, as I will discuss shortly. Because we are relational beings, we cannot and do not need to learn everything on our own. But in order to learn, we should not be required to put our complete trust in any text, tradition, or person. In order to teach, we do not need to ask for perfect trust or devotion to ourselves, to a text or teaching, or to a tradition. All spiritual texts, traditions, and persons at their best express a portion of the truth from a particular perspective. At their worst, they tell us to do things that we know to be wrong and make it seem that unjust social relationships such as slavery or the inequality of peoples, of races, or of genders are "the will of God." Rather than pretending that this is not true, it might be better to admit it from the outset. If spiritual traditions were infallible, then they would never legitimate injustice. Yet they have done so, not once but many times.

If spiritual authorities are not infallible, then we should never be asked to submit our will or intellect to them. "Trust no authorities" was a slogan of the counterculture of the 1960s. According to process philosophy, those who espoused this view were on the right track. In response, it might be said that only abuses of power are wrong. For example, most of us would agree that it is wrong for priests to seduce altar boys and women who come for counseling. Or that it is wrong for Orthodox Jewish authorities to refuse to grant a divorce to a woman whose husband has deserted her. Or that it is wrong for Christians to support segregated schools. Or that it is wrong for Muslim leaders to teach that dying while committing terrorist acts against the enemies of Islam will guarantee a place in heaven. Abuses of power can be corrected, traditionalists argue, within a system that espouses infallible revelation. Yet the question remains: How can we question the abuse of power if we have submitted our intellect or will to authority?

The potential for abuse of power and authority also exists among those who question traditional authorities. A well-known invocation called "The Charge of the Goddess" concludes with a warning: "for if that which you seek, you find not within yourself, you will never find it without."[3] Written and repeated in the context of a world shaped by biblical and priestly authorities, this prayer tells us to trust no authorities in matters of the spirit. Yet ironically the woman who crafted these words collaborated for many years in the fiction that this prayer is ancient and therefore authoritative.[4] Today many who come into the Goddess movement are taught that this prayer was revealed by the Goddess or handed down in secret from the pre-Christian past.

Those who have left western traditions and wish to explore non-western spiritual practices such as yoga or meditation are sometimes asked to submit will or intellect in order to receive teaching. I have a friend who proudly tells others that she made 100,000 prostrations to the "dharma" or Buddhist teaching, before being taught the higher levels of Buddhist practice. She explains that she finds nothing wrong in prostrating herself to the dharma because so far Buddhism has not failed her as Christianity did. Another friend is learning yoga in an ashram where the swami asks for perfect devotion to his person and teachings. She has gained a great deal from the practice of yoga but does not agree with the fundamental premise of the swami's teaching that that the purpose of yoga is to transcend the body. When I asked her why she didn't choose another teacher, she replied that there is a spiritual feeling in his community, and that so far she has not seen any evidence that her teacher has abused power. There are rumors that he has, but she has chosen not to try to find out what is behind them. My friends are typical of others who have left western religions in part because of questions about infallible authority. They would probably never consider submitting their intellect or will to the Bible or to a priest, a minister, or a rabbi. But when asked to give up their personal power and judgment in order to gain enlightenment from a non-western tradition, they are less critical.

Having lived in a foreign country for many years, I understand the desire to give up habits of judgment formed in one's own culture in order to enter into another. Yet I also know that the same kinds of power issues that exist in our culture are found in others. What this means is that we need to be careful about giving up our judgment in any context. In response, it is often said that all religious initiations or conversions require giving up selfish, egotistic, or controlling individual will and judgment in order to open to a wider perspective called "enlightenment," "the love of God," or "union with the divine power." Process philosophy agrees that a selfish and egotistic individual self is not the authentic self, as we have seen in previous chapters. For process philosophy, an enlightened or spiritual person would most definitely be aware that she or he is embodied and embedded in a web of relationships with other individuals, human and nonhuman, and supported and sustained by the love and understanding of Goddess/God.

In the process of my spiritual transformation, which I have sometimes called an "initiation,"[5] I had to give up certain forms of ego control in order to open myself to the power of love that was all around me. For example, I thought that I had to make everything absolutely perfect for the women who came on the first Goddess pilgrimage to Crete and I was afraid of failing them. Part way through the tour I lost my voice. This meant that I could not give all the lectures that I had planned. I was forced to accept help from others when I was supposed to be helping them. Despite my losing my voice and giving up some degree of control, the pilgrimage was a great success. When I got home, I felt the need to stay in bed for more than a week. I wanted to get on with my life, but I couldn't. My body was telling me to listen to it. The message I finally got was that I do not have to control everything in my life and the lives of others in order for everything to be all right. As I listened to my body, I began to give up a relatively more rigid and controlling ego and replace it with a relatively more flexible, trusting, and open one.[6] This did not happen overnight. The process of giving up control is ongoing.

"So," you might be thinking, "isn't that what 100,000 prostrations to the dharma, devotion to a guru, or faith in God is all about? Giving up your selfish and egotistical self?" My response is yes and no. My yes is that I agree that giving up an egotistical or controlling self is the goal of many spiritual paths. When power over is the only model of power we have, then it might seem that the only way to transform an egotistical or controlling self is to "give it up," "give it over," or "submit it to a higher power."[7] My no is based on personal and shared experience of the abuse of power in family, school, university, church, and society. The habit and willingness of individuals to submit to religious and political leaders have been the source of great harm and suffering in our time and others. Military systems, fascist regimes, and many traditional religions depend on the willingness of people to submit to a higher authority. Sexual abuse is rampant in families, schools, and religious institutions. Within systems that appeal to infallible revelation or infallible persons, the notion that we should give up or submit our will or intellect is viewed as acceptable. For those who are lucky enough to have spiritual teachers who do not flagrantly abuse power, it may seem that it does no harm to submit to them. Yet in so doing, we accept the premise that giving up our power in exchange for something is a good thing. No matter what may be offered, the potential for abuse is always there when we give up our personal power. I suggest that submission of the will or intellect is not necessary in the quest for spiritual transformation. We must construct new models in which the giving up of rigid and inflexible or selfish ego control is understood on the model of transformation and power with rather than submission and power over.

From a feminist perspective, there is another problem inherent in the idea that we must give up intellect, judgment, or personal power in order to attain spiritual transformation or insight. Many years ago feminist theologian Valerie Saiving pointed out that the traditional definition of sin as prideful self-assertion, and grace as selflessness or self-giving love, is not necessarily helpful to women.[8] She suggested that the model of spiritual transformation as renouncing ego might make sense for many men who already have a strong sense of self. But for women, the problem is more likely to be valuing the self too little. For women a spiritual path may involve learning self-love rather than selflessness, and spiritual transformation may lead to self-assertion rather than self-negation. While this analysis was offered in a Christian context over forty years ago, it remains valid today. We might want to guard against the (essentialist) interpretation that (all) men are egotistical, while (all) women are self-effacing, but we can recognize that in most cultures women have weaker egos than men do. Women need to learn to trust their own authority, not give it up to another. Process philosophy can also help us to see that even for men the model of giving up egotism should not be confused with submission of the will to a higher authority.

Recognizing the potential for great abuse that is inherent in authoritarian systems, I would never participate in any spiritual or religious tradition that asked anyone to submit intellect or will to any allegedly higher authority. Some may choose to situate themselves within such traditions while openly challenging claims to infallible authority or revelation. Challenging claims to authority is another way to deconstruct widespread habits of submission. But those who challenge authority too insistently may find themselves thrown out. Here I think of former Roman Catholic priest and theologian Matthew Fox, former Iyengar yoga teacher Angela Farmer, former devotee of Guru Ma, Andrew Harvey, and the Protestant women who lost their jobs as a result of attending the first Re-Imagining Conference. They learned the hard way that traditions they were following would not allow them to continue questioning allegedly infallible truths. This is an important lesson.

"All well and good," you might be thinking, "for those who have the power to challenge all authorities to do so. But what about the oppressed for whom the word of God spoken on their behalf is liberating?" One of the ways that the oppressed have challenged entrenched power is by claiming access to their own understanding of infallible revelation. When Martin Luther King, Jr. preached against white racism, he did do so in the name of God. African American religion is not so much about deconstructing infallible revelation as it is about offering another interpretation of its meaning. Liberation theologians speak of God's preferential option for the poor, while some feminists speak of God's preferential option for poor women. At times these claims are made in the name of revealed truth or infallible revelation. Some would argue that when the oppressed speak truth to power, they are not helped by theories of the ambiguity and uncertainty of all truth claims. Indeed it might be argued that in relativizing all truth claims, deconstructionism allowed academics to dismiss challenges to the elite white male power by labeling them "unsophisticated theory." There is some truth in this charge.

Recognizing that theories of the uncertainty and ambiguity of truth can be used to stifle dialogue about power alerts us to the potential for the abuse of all theories. If too much power is held by any group, that group can probably turn any theory to its own advantage. This reminds us that having the right belief or theory is no guarantee that we will think or act ethically. On the other hand, the notion that it is possible for any group to have access to infallible truth has done so much harm in the world that I find it hard to imagine that anyone would want to defend it, even in the name of the liberation of oppressed groups. The illusion of being infallibly correct has allowed those with power to feel justified in torturing and killing alleged infidels. Wars have been and are being fought "with God on our side." Innocent women and children have been killed with "God on our side." The illusion of having God on one's side leads to the demonizing of others. Thus their deaths can be justified. When the oppressed take power with God on their side, they often carry the habit of demonizing others with them, as is all too well known. Do any of us really want to promote this habit of thinking?

But if we cannot trust spiritual authorities, then how do we know? For some traditional western philosophers, reason took the place of God. This was a step in the right direction, as the ability to reason is in principle available to all human beings and not only to an elite group. However, following Plato, philosophers tended to imagine that truth exists in an immortal and unchanging realm. Human beings could have access to unchanging truth through reason purified of emotion and connection to the physical world. Since the capacity for "pure reason" was imagined to exist in a (male) mind separated as much as possible from the body and relationships, philosophers tried to separate themselves from the concerns of everyday life. Because it recorded the efforts of men to think about "eternal truth," philosophical tradition gained a quasi-religious status, and for some, the philosopher became a kind of priest. The fact that most philosophers write in a self-referential language filled with technical distinctions made by themselves and other philosophers that even otherwise well-educated people find difficult to understand contributes to this mystification. Sigmund Freud questioned the power of conscious reason, asserting that the unconscious plays a far greater role in human life than had been recognized. Karl Marx showed that western philosophical traditions express unrecognized class interests. Yet both continued the tradition of writing in language that was not easily understood. In different ways, both Marx and Freud were viewed as priests offering a new truth, and their works became sacred texts to be interpreted by their followers. Blacks, Hispanics, Native Americans, Asians, colonized peoples, women, and homosexuals extended the criticisms of Marx and Freud, pointing out that what had been touted as the fruits of pure reason was in fact the thinking of elite white heterosexual men educated in European traditions. Deconstruction continued this line of criticism, alleging that there can be no access to universal or unified truth and that all claims to knowledge of "absolute" or "universal" truth disguise a will to power. While claiming to demystify the power inherent in the control of knowledge and knowing, deconstructionists write in a technical jargon that "is difficult even for an ordinary person with a Ph.D. to understand," as one of my friends once said in frustration.[9]

Whitehead and Hartshorne departed from western philosophical tradition's emphasis on eternal truths when they asserted that we think through the body in connection to other individuals in a changing world. As Whitehead said, "In philosophical discussion, the merest hint of dogmatic certainty as to finality of statement is an exhibition of folly."[10] Process philosophy is in principle open to the criticisms of traditional western philosophical thinking posed by Freud and Marx, and by African Americans, Hispanics, Asians, colonized peoples, women, homosexuals, and deconstructionists. If all thinking comes through the body, then there can be no infallible truth in philosophy. There can only be approximations inherently limited by class, race, power, gender, sexual, historical, and other perspectives. Whitehead and Hartshorne recognized at least some of this. This too is a step in the right direction.

Yet as noted in the introduction to this book, both Whitehead and Hartshorne were heterosexual white men who came from families with intellectual privilege. They felt comfortable situating their work within western philosophical tradition. They took it for granted that for westerners at least, access to (limited and fragmentary) philosophical truth would be found through intensive study of and dialogue with the "great thinkers" of philosophical tradition (almost all of them white and male and privileged and not openly homosexual). Thus, though both men sympathized with feminism, women thinkers are rarely considered in their works. Nor did either ask whether the western tradition is seriously flawed because of the exclusion of women's and other voices. In my opinion, Whitehead and Hartshorne failed to question the authority granted to western philosophical tradition as strongly as they might have, given that they understood the mind to be situated within the body and the changing world of relationships, both personal and social. This is an area where feminists and others can extend and clarify the implications of process thinking. Feminists, especially feminists of color, have a much more radical understanding of the limitations of elite white male traditions of thought. We understand the complex ways in which the body and gender shape understanding. We are more open to finding truth in the voices and experiences of those who have not usually written philosophy. This book is part of a larger feminist effort to reconstruct "pure reason" as "embodied, embedded knowing."[11] If there is no source of infallible revelation, then in one way or another we must trust in our own powers of knowing to discover the meaning of our lives.

The question remains: What is embodied, embedded knowing? This question is as complicated as the question of the self. Given the individualist tendencies in the West, it would be easy to imagine that all authority is to be located in a radically autonomous rational self that is not necessarily connected to its own body, to others, or to a social world. Yet for process philosophy, the self is not an isolated rational individual. For process philosophy, the self is always a self-in-process. This means that our knowledge will be deeper and fuller when we are open and flexible rather than rigid and closed. For process philosophy, the self is also embodied. We know by listening to our bodies and taking account of our feelings. For process philosophy, the self is also a self-in-relation, embedded in the world. We know by listening to and learning from other individuals, human and nonhuman. We know by opening ourselves to animals, plants, or the cells of plants, minerals, and to atoms and the particles of atoms. We know by sharing insights with others, by expanding our perspectives through conversation and study. For process philosophy, we are also related to divine power. We know by opening ourselves to the persuasions offered by Goddess/God. For process philosophy, human beings are also thinking individuals. We know by observing and reflecting upon the world in which we are embedded. Deeply involved in a changing world, we seek ever-changing and deeper understanding, not access to unchanging or eternal truth.

What about teachers? In life we will all have teachers of various kinds, beginning in our families. Learning from others is part of life. The question that most of us learn through experience to ask is whether our teachers themselves are open to changing life in all of its variety. Do our teachers, including our spiritual teachers, delight in difference and in our growing creativity and freedom? Or are they more interested in replicating themselves? Do they believe there are many paths or only one? In my own experience I have been deeply disappointed by my parents and by teachers who did not always delight in my growing creativity and freedom. My parents appreciated my artistic talent and my intelligence until I made the decision to pursue a graduate degree. They may have feared that my desire for what they perceived as "too much" education threatened their control over me. And indeed it did. I did not settle down and raise a family as they had hoped. I have not voted for the political party they supported. I have feminist ideas that challenged the compromises they made in their lives. In college I had two professors who opened new worlds of understanding to me and encouraged me to pursue graduate education. Yet at a certain point, these professors also ceased to support me and my work. One of them told a colleague that he was disappointed that I had "wasted [my] brilliant mind on all that feminist stuff." The other encouraged my feminism until it forced me to leave the church. To be fair, my parents and teachers probably thought that they really did know what was best for me. The pain caused by the failure of my parents and teachers to support my growth was exacerbated by my own unrealistic expectations of them. I did not understand them to be limited and flawed individuals. I had expected them to be "infallible," their support of me unwavering. A process view suggests that good parents and teachers would be open to their childrens' and students' freedom and creativity, and that they would teach children and students that no authority, including their own, should ever be perceived as infallible.

What about spiritual and religious traditions? Many of us will also learn from spiritual traditions, western and non-western, traditional and countercultural. In the past, people were born into a tradition, accepted its teachings, and followed the way of life it set out, to a greater or lesser degree. Most traditions had religious rebels and many of these (including some of my ancestors) fled to other countries, including America. What is different about the contemporary situation in all countries, especially but not only in North America, is increasing religious pluralism. My relatives have included Roman Catholics, Christian Scientists, and Methodists; my family went to a Presbyterian church; one of my brothers is a Mormon; one of my cousins followed a Hindu guru; another trusted a psychic; some of my ancestors were Quakers. My closest friends have included Protestants, Catholics, Christian Orthodox, and Jews, as well as atheists and agnostics, Native Americans, Goddess feminists, and followers of Buddhist and Hindu traditions. In my own life I have practiced Presbyterianism, Roman Catholicism, feminist Goddess spirituality, and my own eclectic version of Greek devotion to the "Panagia," the female divine in the person of Mary, and the female saints. I have learned a great deal about the relation of body and spirit from the practice of gestalt therapy, yoga, and various traditions of healing with the hands. Feminism came into my life with the force of revelation. I have also been deeply influenced by both Judaism and Native American spirituality through friendships and reading. For me nature has always been a teacher.

In the situation of pluralism, some, like my brother Kirk who is a Mormon bishop, seek the one true way. But for many of the rest of us, pluralism means that we are open to learning about the divine power in different contexts and through a variety of methods. We may feel more committed to one tradition than to another—as I do to Goddess feminism and my friend Judith does to Judaism. But we are likely to attribute our being in one tradition rather than another to choice and chance rather than to our tradition's infallible authority. In addition, if we are honest and our minds are flexible, we will be able to acknowledge and relish having learned some of the truths of our lives in traditions other than our primary one.

Can the self be its own authority? Should we put our faith in insights that come from the deep self? When we think of the self as a spiritual authority, we often speak of what some have called the "deep" self. Most of us have a sense of what this means. We can think of times in our lives when everything seemed clear and vivid. In those moments it felt as if we knew the meaning of life. Sometimes we connect with the deep self in therapy, meditation, prayer, yoga, ritual, dream-work, or trance. A feeling of wholeness may come over us while riding a horse or when walking by a lake. Some of us hear voices or see visions. Others gain insight in therapy, in conversation, or while reading or writing in journals. In my life there have been a number of moments of spiritual clarity. I have already mentioned the revelations that came when I argued with God and Goddess and as my mother died. There have been other times of dramatic revelation in my life.

When we turn away from infallible authorities outside the self, we sometimes imagine that we can find an infallible authority within the self. For some Hindus the deep self is divine. For some Buddhists, meditation erases all boundaries between the self and the divine. Some in the Goddess movement say that all women are Goddesses. Others think of insights that come in trance, meditation, dreams, visions, or divination as revelations from the Goddess. Jungian psychology says that divine wisdom is reflected in our dreams. In western culture we have more experience with rational as opposed to non-rational ways of knowing. Beginning to open ourselves to other than rational ways of knowing is especially important for us. Yet from a process point of view, it is important to remember that whatever we experience in our deepest selves is also fragmentary and limited. When it comes down to it, all of our insights are our insights, experienced in our bodies, shaped and limited by our pasts, the present, and our hopes for the future. There can be no final authority and no absolute certainty within the self just as there is none without. This means that information that comes to us from other than rational sources about the divine presence in our lives must also be evaluated and reflected upon both individually and in community with others. When we do not expect certainty or infallibility from any source, not even from our inner selves, then we can truly open ourselves to changing life in changing bodies embedded in the world.

So how do we know the presence of Goddess/God in our lives? Though there is no absolute certainty or infallible truth in our lives, there can be a kind of open and flexible inner knowing of the divine presence that arises through the body, experience, relationship, community, and reflection. For me this is not the result of the infallibility of any particular text, teacher, tradition, or of any infallible experience or insight I have had. Rather this is woven out of patterns discovered in a wide variety of experiences and relationships in my life. My Grammy's love. The peacocks in her garden. My Nannie's faith that her prayers would be heard. Candles and rosary beads. Tell me why the ivy twines. Red and yellow, black and white, all are precious in His sight. The waves crashing against my body. My brother Kirk's birth. My brother Alan's death. My grandparents' deaths. The Hebrew Bible. Buber, Heschel, and Wiesel. The Vietnam War. The Holocaust. Racism. Sexism. In God is a woman like yourself. She shares your suffering. Lessing, Shange, Atwood, Chopin, and Rich. We all come from the Goddess and to Her we shall return. The golden laughter of Aphrodite. Love filling the room. My mother's death. The darkness of the cave. Giving up control. Scops owls calling all night long. The dogs and cat I fed this morning. Yoga energizing my body after my bath. Calls from Cristina and Nancy. Writing with joy. Maro greeting me as she passes my door. "My lovely sister" Madhu in Nepal studying to be a nurse.[12] Other girls in her country being sold into indentured servitude. Light and darkness.

What follows is a series of notes about my changing awareness of and response to the divine presence in my life. What I say here comes through my body and my experience, shaped by my history and my desires, in conversation with my friends and those whose worlds I have shared, and in relation to Goddess/God whose presence shapes my life. What I say can and must be expanded and extended from other perspectives. Yet the only way I know to begin is to talk about how I know and reflect on my own embodied, embedded knowing. I do not begin with proofs for the existence of divine power, but rather speak of my awareness of it.

The presence of God was taken for granted in my childhood. My grandmothers prayed daily for God's help and guidance, and in Sunday school I learned that God is love. I felt the presence of God in my mother's mother's garden and while swimming in the sea, as well as in my father's mother's fervent prayers for my uncle Bobby, who was troubled. My brother Kirk was born when I was ten years old. He was a miracle to me. I gave him all the love I had within me. When my brother Alan died a few years later, only five days after he was born, I was introduced to death. I had prayed with every fiber of my being that he would live, but my prayers were not answered. My father's mother and my mother's father died that same year. From that day on I knew that the love of God had to include the tragedy of death. Trying to understand that mystery propelled me to study the Hebrew Bible and theology.

Some years later, when I became aware of masculine and militaristic and anti-Judaic language for God in Christian hymns, prayer, and liturgy, and in the Bible, my body began to tell me that I could no longer participate in the Christian tradition that had once been nourishing for me. Whenever I attended a church service, my stomach would clench and I would feel as if I wanted to throw up.[13] When I realized I could no longer call myself a Christian, I was cast back onto my inner sense of knowing. Most of the theology I had studied no longer made much sense. I did not have any mentors who supported my new insights, and my friends were floundering as I was, though on somewhat different paths. For a long time I felt very alone. I began thinking again about my experiences of the spirit in nature, sexuality, and community. I didn't always have a language to express what I felt. In time other women "heard me into speech"[14] in conversations and in groups we formed with the intention of changing our lives and the world in which we lived.[15]

During this period I found the writings of Doris Lessing, Ntozake Shange, Margaret Atwood, Adrienne Rich, Kate Chopin, and Denise Levertov to be like sacred texts that affirmed my growing awareness that women's spiritual quest had its own themes and rhythms. My students and some of my friends also responded strongly to these writers. When I began speaking and writing about women's spiritual quest and eventually published Diving Deep and Surfacing,[16] I found that other women felt I was speaking for them. I have never felt as alone since. Doris Lessing's insight that whatever is happening at the deepest level of the self is not only personal because we are all shaped by the same social forces gave me courage to continue to speak my deepest truths.[17] From these experiences, I learned to trust the messages that came through my body, to seek the words of women writers, and to hope that in conversations with women students and friends we could together find a way out of the labyrinth. I was developing a way of thinking through the body that involved experience and reflection and was embedded in community.

In those years, I was also searching for the God who is a woman like myself. I felt her presence the night I expressed my anger at God, as discussed in chapter three. By chance, Naomi Goldenberg discovered a workshop offered by a then-unknown woman named Starhawk on witchcraft and suggested that we attend it. Hearing Starhawk speak about the Goddess, I felt that I had come home. I began speaking about the Goddess to my feminist theological friends, naively assuming that they would feel as I did. This did not happen. Several years later two friends and I formed a women's ritual group, which we sometimes called a coven, to explore Goddess spirituality through ritual. This community supported my growing relationship to the Goddess as my academic friends could not. Though we often relied on recipes for ritual found in books by Starhawk and Z Budapest,[18] we also trusted our own intuitions and referred to other traditions. None of us ever felt we needed to follow a particular Wiccan tradition, although we sometimes gleefully called ourselves "witches" and in deep seriousness initiated each other as "priestesses" of the Goddess.[19] Over the years, those rituals that were most closely associated with Wiccan tradition stopped working for me. I liked creating a circle and calling in the powers of the four directions. I liked working with energy and healing with the hands. But I never found the need to use a ritual knife or understood what the "Guardians of the Watchtowers" were.[20] Secrecy seemed unimportant to me. Our group never felt nudity to be essential. The words "witch" and "priestess" began to seem less descriptive of my path, especially as I spent more and more time in Greece. I began calling myself a participant in the Goddess and women's spirituality movements. I felt more comfortable with less elaborately constructed rituals and was less attracted to magic, spellcasting, and trance. Later I learned that most of the parts of Wiccan tradition I no longer needed were in fact adaptations of Masonic and Rosicrucian rituals.[21] These experiences reinforced the importance of community and taught me to look at all traditions, including those alleged to be feminist, with a double eye. I learned from Wiccan tradition, yet I never submitted my will or judgment to it. In the same way, I could learn from other traditions while not accepting the claim of any tradition to have access to the one true way. During this time I also was expressing my love for the world and my understanding that we must protect the earth that is her body through anti-nuclear activism and the creation of a slide show called "Genesis/Genocide: Women for Peace." Thinking about the possibility of nuclear war raised questions about the nature of divine power that I struggled with during that time, as I discussed in the previous chapter.

When I taught in Greece in the summers of the 1980s, I created rituals with other women that celebrated women's bodies and sexual longing at Aphrodite's temple in Lesbos. During one of the rituals I heard a golden laughter rise above the ruins. It seemed to affirm my desire for sexual love and ecstasy.[22] Another time I felt that the Goddess Aphrodite had answered my prayer when a lover synchronistically arrived in my life soon after a ritual at the temple. At first ecstatic and joyful, this relationship was also destructive. Its ending plunged me into a depression that lasted for several years. As I later realized, my despair was not only a response to the loss of a relationship I desperately wanted, it was also a crisis of faith. When I first started following the Goddess path, I naively believed that if I listened to my body and what seemed to be my deepest intuitions, I would be following the will of the Goddess. When my lover disappeared, I felt that I had been betrayed not only by him but also by my deepest intuitions and by the Goddess.[23] The question of what kind of trust to put in messages that seemed to come from my deepest self was not easily solved and continued to haunt me. Over time, I have come to understand that I had made the mistake of imagining that insights that came from other than rational sources came directly from the Goddess. I now recognize that while insights that come from what we call the deepest self can and do guide us, we must also speak about them with others and reflect on them, weighing their value in terms of how they make sense of our lives and life as a whole. The insight I had when my mother died that we are surrounded by a great matrix of love has stood the test of time. It has been confirmed as I open my heart more and more to the world and the individuals in it.

For me all of these experiences and many others have taken shape as a now-familiar sense of inner knowing that I am alive, embodied, and embedded in a changing web of life and relationships co-created by myself, other individuals, and Goddess/God whose love supports and sustains the world. As my knowing becomes deeper and stronger, I depend less on extraordinary experiences to guide me, because I begin to find the divine presence and love in everything. Insights are less dramatic but more frequent. Feeling that my body and soul are connected to the sky and salty water when I swim far out into the sea. Sensing that I am not alone when I light a candle in a church and look into the liquid brown eyes of the Panagia. Trusting the darkness when I descend into the Skoteino cave. Knowing that it is right to try to save the lives of girls in Nepal and the wetlands of Lesbos.

Throughout this chapter I have asked: How do we know? And: How do we know the divine presence in our lives? Reflecting on these questions again, I answer that we know in the living of our lives. No one else can know for any one of us. No one else can know what we know. We listen to our bodies and the feelings that come through our bodies. We share our lives with others in friendship and community. We read. We pray or meditate. We reflect. And yet there is much we do not know. Life is a process of change, and we know only small parts of it.

Where does process philosophy fit into all of this? As we go about the process of living, we reflect on its meaning. Process philosophy can help us think about our experiences of the world. The usefulness of process philosophy for me or for you is its ability to provide a framework that can help us make sense of past and present experiences and to open us to a richer and fuller experience of life in the future. This book is a demonstration of how this works. Yet process philosophy as expressed by Whitehead and Hartshorne or by me is not an absolute truth. As I have found weaknesses as well as strengths in the thinking of Whitehead and Hartshorne, so will you find strengths and weaknesses in mine. Because it does not claim finality, process philosophy is inherently open to expansion, correction, or transformation from other points of view and from new insights and knowledge. Philosophy is a dialogue or conversation about the meaning of changing life. Lack of certainty is part of what it means to be open to life in all its complexity.

For process thinking there are no absolute truths in life, in religion, or in philosophy. All knowing is embodied and embedded knowing. All knowing is partial and fragmentary. All knowing is relative. Yet there are relative truths, learned by being open to the body and the feelings that come through the body; to other individuals, including human beings, animals, plants or cells of plants, and atoms and particles of atoms; and to the persuasions offered by Goddess/God. Relative truths help us to make sense of the depth and breadth of our experiences. Relative truths help to enrich life for ourselves and others. Understanding relative truths is a process of embodied, embedded knowing. Yet if all truth is relative, then why have so many individuals, religions, and philosophies sought or claimed access to infallible truth? Could it be that this common theological mistake, the desire for certainty or infallible truth, is based in an inability to accept limitation and vulnerability that also begins with a rejection of the female body?

Chapter Seven

Reason for Hope

For process philosophy, the world is co-created by Goddess/God and all other individuals, including particles of atoms, cells, animals, and human beings. Whitehead described the process of life as an adventure. He felt that novelty and surprise make life interesting. The open-endedness of life provides opportunities for the exercise of creative freedom, which gives life meaning. And yet, as many of us are all too well aware, the exercise of creative freedom by human beings (and perhaps to a lesser extent by other individuals) has not always been positive. Indeed, the word "adventure" seems inappropriate to describe seas and lands poisoned by bombs and chemicals; species extinct or struggling to survive; people fleeing racially and ethnically inspired violence; women and children forced to endure rape, genital mutilation, and incest; the threat of nuclear annihilation. Whatever their flaws and whatever the intellectual problems they create, traditional religions at least offer the hope that in some time or place suffering will end. Can process philosophy offer reason for hope to those who are suffering and to those who feel for the suffering of others?[1]

For traditional theologies, faith in an omnipotent and omniscient God is the reason for hope. Classical theism solves the problem of the limitation of human knowledge about the future with the doctrine of God's omniscience. If God is all-powerful or omnipotent, it is said, then he must also be all-knowing or omniscient. Traditional theologies assert that God knows the past, the present, and the future with perfect clarity. If God knew only the past and the present, then his knowledge would not be complete, and therefore it would not be perfect. Yet God is perfect; therefore, he must know the future as well as the present and past. From this perspective, many theologians have defined hope as "belief in things not seen." We have not seen the future, but God has seen it and knows it with unfailing knowledge. Since the future has been or will be created by an all-powerful God, we can rest assured that it is and must be a good and just future, worthy of our hope.

For process philosophy, as discussed in chapter four, the doctrine of divine omnipotence is not compatible with the divine relativity or relatedness or with the freedom of other individuals. Hartshorne also called the doctrine of divine omniscience a theological mistake. It is based on the false premise that the future has been or will be created exclusively by an omnipotent God. For process philosophy the future cannot be known by anyone—not even by Goddess/God—because it has not yet been created. If individuals other than Goddess/God really do have creative freedom, then the actions, inactions, and reactions of all them will create the future. Goddess/God is fully involved with other living individuals in the process of co-creating the future out of the present and the past. Process philosophy asserts that no individual—not even Goddess/God—can step outside of the creative process that occurs in time and space in order to see what will happen in the future.

There is, however, another and more limited sense in which Goddess/God can be said to be omniscient. As we have seen in previous chapters, Goddess/God is omnipresent to all individuals who live or have ever lived, and the world is the body of Goddess/God. From this perspective, Goddess/God has perfectly accurate and perfectly sympathetic knowledge of the present and the past. Moreover, insofar as the future is constructed from the present and the past, Goddess/God can be said to have a clearer and deeper sense of the possibilities of the future than any other individual. However, even Goddess/God can be and will be surprised by the future that actually is co-created.

If Goddess/God cannot have full and unfailing knowledge of the as-yet-uncreated future, this means that the fate of the earth and, for that matter, the universe cannot be known—not even by Goddess/God. Will the adventure that is life on earth end in tragedy? This is a distinct possibility. Will human beings individually and collectively recognize that we are connected with each other and all beings in the web of life? Will we learn to use our intelligence and creative freedom with compassion for each other and all beings in the web of life? Will we stop killing each other and destroying the environment before it is too late? This is possible. But it is not guaranteed.

For process philosophy, the reason for hope is the creative process of the universe itself. This process is supported and sustained by Goddess/God. Yet in asserting that the freedom to co-create (and possibly co-destroy) the conditions of life on this planet is partly in the hands of individuals other than Goddess/God, process philosophy, more than many other theologies and religious philosophies, recognizes the creative (and destructive) power of individuals other than Goddess/God. Process philosophy thus acknowledges the potentially tragic character of life on earth. In our time the tragedy of suffering and unrealized potential is a possibility not only for individuals but also for the evolutionary process of the planet. The fate of the earth at this moment in time is to a large extent in the hands of its most free and creative individuals, human beings. In many respects this is a new situation. "It took hundreds of millions of years to produce the life that now inhabits the earth—eons of time in which that developing and evolving and diversifying life reached a state of adjustment and balance with its surroundings. . . . Only within the moment of time represented by the present [twentieth] century has one species—man—acquired significant power to alter the nature of his world."[2]

Through our technologies, human beings have gained an awesome power. At first this seemed to many to be the ability to "control" nature. Those who first asserted that human beings could control nature held two contradictory views about nature. On the one hand, they followed modern science and said that "nature" was "mere matter" that could be shaped by the rational powers of the human mind. On the other hand, building on long-standing theological traditions, they sometimes spoke of "Nature" (often personified as feminine) to be a "power" separate from "man," intent on thwarting the project of human life on earth by causing the "natural evils" of death, decay, droughts, hurricanes, earthquakes, and the like. Yet they hoped that rational men could "tame" or "subdue" her power. Process philosophers, ecologists, ecofeminists, systems theorists, and others are beginning to take a more holistic view of the web of life. We understand that human beings do not stand outside nature or the web of life but are part of it. "We are nature. We are nature seeing nature. We are nature with a concept of nature."[3] For holistic thinkers, "nature" is not mere matter but a complex system of living beings including human beings that has developed over millions and millions of years of evolution. Human beings, who are part of nature, will never be able to control nature. Yet, tragically, human beings now may have the power to destroy the web of interconnection that has made life, including human life, a possibility on planet earth. What human beings do with our creative freedom in the twenty-first century will have enormous consequences for the future of life on earth not only for human beings but for all other species, and not only for the living but for the generations to come. This is an awesome challenge and responsibility.

Process philosophy has a very different understanding of human responsibility for the fate of the planet earth than many other theologies and philosophies. When the focus is on divine omnipotence and omniscience, traditional religions may encourage passivity in the face of suffering. Some believers imagine that the divine power will save the earth at the end of time no matter what we do, or that it will offer us another life in heaven or on earth. Others focus their energies on getting on the good side of omnipotent power so that they will be included in the salvation or redemption it will bring about. They may attribute the injustices in this world to the will of God and therefore find no compelling reason to do anything to alleviate them. Or they may believe that because God will reward those who suffer in this life in heaven or another life, there is no urgent need to struggle to create a more just world. Of course traditional religions have also encouraged works of charity and struggles against injustice. Yet if doctrines of divine omnipotence and omniscience are retained, the need for humans to struggle to end injustice may seem less compelling. Because it denies omnipotence and omniscience, process philosophy places the responsibility for ending humanly created suffering squarely in human hands. But unlike atheism or humanism, process philosophy believes that human beings are supported and sustained in their efforts to enhance the flourishing of life by a divine power that profoundly desires that all individuals enjoy life as much as possible.

It can be very upsetting to realize that there are no guarantees that life will continue on this earth in richness and diversity. What if all of our efforts to save the world come to nothing? What if the world is blown up in a nuclear holocaust? What if we continue to poison the earth to the point that human beings and most other species cannot live or can survive only in weakened forms? When I look at the world in which we live, it has often seemed to me that pessimism about the future is the most sober and realistic conclusion. Over half a century ago, Albert Schweitzer said, "Man has lost the capacity to foresee and to forestall. He will end by destroying the earth."[4] Optimists often hold the naïve view that all human development is progress and see all technological innovations as unqualified advances in civilization. They forget that more technologically advanced European colonists in Africa and the Americas destroyed native cultures that lived in greater ecological balance with the planet. Or that modern railroad systems enabled the Nazis to deport Jews from all over Europe to the death camps. Or that the weapons of war are now more deadly and in the hands of children. Or that one person with a finger on the trigger can set off a nuclear war. Or that every nuclear power plant that is constructed contains within it the possibility of another Chernobyl. Or that we are poisoning the earth with pesticides, residues of weapons, and chemicals, making it unfit for life. In time we might be able to make progress in overcoming the historic injustices caused by racism, sexism, heterosexism, ethnic, and religious hatred. In time we might even be able to end war. But if we continue to destroy the earth, will we have the time? At the close of the twentieth century, animal biologist, chimpanzee specialist, and environmental activist Jane Goodall warned us again, "I do have hope for the future—for our future," she wrote. "But only if changes are made in the way we live—and made quickly."[5] The thought that human beings may destroy the possibility of life on earth not only for ourselves but for others can be paralyzing. It has paralyzed me.

A review of Jane Goodall's spiritual autobiography *Reason for Hope* provoked a quantum shift within me, releasing the paralysis I had felt about the fate of the earth. Goodall's book taught reviewer Katy Payne that "hope . . . is not found in optimism so much as in a primal understanding of what matters most."[6] In other words, the reason for hope is not to be found in the knowledge or rational calculation that our efforts will succeed in saving life on earth but rather in the conviction or inner knowing that it is right to try. This thought (which I did not find expressed in exactly the same way in Goodall's book) was profoundly liberating for me. Like many others, I have spent quite a bit of time wondering whether anything I can do could make enough difference to save the earth. In light of the enormity and different kinds of problems we face, no action that I could imagine taking seemed like it would make a difference. What I learned is that I do not have to know whether or not my efforts combined with those of others will actually end up saving the earth. What I do need to know is that it feels profoundly right to me to make whatever efforts I can to help others and to protect life. Since then, the energy I once wasted in trying to know what (from a process point of view) cannot be known anyway—the future—has been freed up to do what I can do. Whether it is writing this book, or sending money to support Madhu in Nepal, or giving money to the Global Fund for Women and the World Wide Fund for Nature, or helping to form the Friends of Green Lesbos, I no longer ask: "Will it be enough?" I feel grateful that I am in a position to do something, and ask instead: "What more can I do?" No one of us can do everything. We all need to find something we can do.

It now seems to me that the question of optimism or pessimism about the fate of the earth is the wrong question. "What," I asked earlier, "if all our efforts to save the earth come to nothing?" The assumption implicit in this question is that if in a hundred or two hundred or five hundred years the earth is so poisoned or degraded that it cannot sustain human life and the diversity of life, then all of our efforts to save life on this planet will have been in vain. But this is not the right question. Even if we knew for certain that in two hundred years there would be absolutely no more life on earth, would it be reasonable to say that all of our efforts to save it were futile? Yes, if the end result is our only concern. But if we look at the process rather than the end result, it makes no sense to say that our efforts to preserve and enhance life come to nothing. If Madhu in Nepal is helped and goes on to help others, that is something. If even one life is saved, that is something. And the truth is that we really cannot know the long-term consequences of any action we take. One small act could be the one that turns the tide.

A number of years ago, a story circulated called "The One-Hundredth Monkey." In this story, the monkeys on a certain island only ate coconut meat when they found a coconut that had split open. Eventually one monkey figured out that by hitting a coconut on a rock it would break. At first this technique was taught by one monkey to the next. But after the one-hundredth monkey learned the technique, it did not have to be taught again.[7] This story can be read as a parable about how common assumptions of cultures can shift. In the nineteenth century those who thought slavery was wrong were a minority. They had to teach others to see that slavery could be ended. Now for the most part this question does not have to be argued. (This does not mean that all slavery has ended or that the effects of slavery have been fully redressed.) Today parents who refuse to let their children play with toy guns are a minority. But at some time in the future, it is possible that the idea that children should be allowed to play with toy guns will be unthinkable because war will no longer be glorified. Today those who use pesticides speak of the "necessity" of doing so. Yet is possible that in the future, people will find it incredible that past generations could have been so short-sighted. Commonly accepted patterns of thinking and action do not change magically. But with concerted educational efforts, change is possible.

For process philosophy the reason for hope is the creative process of life itself. If human beings have created many of the problems that limit and threaten the possibilities of life on this earth, then we have the capacity to solve them as well. The way human beings use our creative freedom may determine the fate of life on earth. "But," you might be thinking, "what can creative freedom possibly mean to those human beings who like rats in the maze have had their options severely limited by others? The child dying of AIDS in Africa? The mother who does not have enough to feed her children in Indonesia? The ten-year old boy who has been conscripted into the army and forced to kill? The woman in Bosnia who was gang-raped by soldiers who once were her neighbors? The twelve year-old girl in Detroit who tries crack cocaine at a party and lets her boyfriend have sex with her? How much creative freedom do these and millions like them have? And what meaning could process philosophy possibly have to them?" Unfortunately there are many people in the world whose freedom is so limited that it might appear to be negligible. It might seem meaningless to speak about creative freedom to them or, worse, insensitive to their real needs.

And yet. Isn't the child dying of AIDS better off if he is cared for and loved until the end of his short life in a hospice created by AIDS activists in his community? Might the woman in Indonesia be better off if protests against the exploitation caused by the spread of global capitalism succeeded? Or if women in her community organized to change the conditions of their labor? Can the ten-year-old boy be rehabilitated in a community of care and learn to tell others about the cruelty of war? What if the raped woman in Bosnia speaks the truth about the horror that was done to her, gaining the courage to join an international movement to designate rape as a war crime? What if the twelve-year-old girl in Detroit finds a program for teenage mothers, finishes school, and gets a job that can help her support the child she had too young?

All of these things can happen and have happened. I am not saying that they happen more often than they do not. Ours is very far from being a fair and just world. What other solution is there but to protest against injustice and to work to change the conditions that lead to great suffering? To help those who have suffered to heal so that they too can begin to experience joy in their own lives and gain the courage to work to change the world for others in their communities? Can my effort or yours end all of the humanly created suffering in the world? No one knows. Can my effort or yours help individuals who are suffering regain the ability to make creative choices in their lives? Without a doubt. Is it worth trying, knowing that we may not succeed in changing the larger picture? Of course it is. Moreover, we do not know what the fruits of our efforts can be, or who may join us. Even Goddess/God does not know the future until it happens. Why not make the most creative use of our freedom that we possibly can and help those whom we can help? I don't see any other choice.

As I have said, our hope is rooted in knowing that it is right to try to save abundant life on planet earth. Yet the conviction that it is right to try to save the earth does not tell us how to go about doing it. What guidance can process philosophy offer us as we co-create the future? In thinking about this question, it is important to recognize what philosophy can and cannot do. Philosophy is a product of reflection on the meaning of life. A philosophy is a description of reality that includes a system of values. Thus, we can look to philosophy to provide a world-view and values that can help us to think about how to save and enhance life on this earth. On the other hand, no philosophy is a full-fledged religious symbol system. A philosophy of religion can provide a context in which we can think about and evaluate the roles played by religious and other symbol systems in creating and helping to solve the problems of the world, as I will discuss in the last chapter. But we cannot expect process philosophy to express and evoke feelings such as sympathy or compassion, outrage or a concern for justice. This is the appropriate role of religious and other symbol systems. Similarly, process philosophy can provide a framework in which we can think about and evaluate various social, economic, scientific, technological, or political theories that have helped to create or alleviate suffering in our time. But no philosophy can be expected to provide the details of the economic, scientific, technological, political, and other solutions to the problems we face.

Process philosophy's world-view is holistic. Human beings are integrally related to the web of life and cannot survive without its support. Individuals other than human beings also have intrinsic value. Process philosophy thus urges us to think about all human projects in relation to the whole web of life, human and nonhuman. For process philosophy the world is socially constructed. The future must be created out of the past and the present. This means that we must take account of a variety of already existing relationships, structures, and institutions as we attempt to change the world. Because of the complex nature of the past and the present, it is unlikely that any problem will ever have a single cause or a single solution. A web of interrelated causes must be addressed. We also need to be aware that as we go about solving one problem, we may inadvertently create or, if we are lucky, inadvertently solve others. Though human beings do have a degree of freedom, we must work with the world as we find it. This means that even if we can imagine a more just and joyful world, we cannot easily change the world to suit our visions. We must always work with others, and we must work in relation to structures of justice and injustice that we ourselves did not create.

Besides offering a holistic world-view and an understanding of the social construction of reality, process philosophy also offers a set of values that can aid us. These include interest, enjoyment, embodiment, freedom, creativity, sympathy (understanding, love, and compassion), relationship, change, and power with rather than power over. As we have seen in previous chapters, these values are not independent but interdependent. Thus, each value must be understood in relation to the others. Freedom can only occur in the context of relationships of some kind, while relationships at their best encourage the development of freedom. Our enjoyment of the world is expanded by having sympathy with the feelings of others, while feeling sympathy with others makes us want to help them to enjoy life more. Relationship, enjoyment, sympathy, and other values occur in bodies. Everything we create is in a process of change. As we think about the world we are co-creating, we can ask if programs and courses of action take account of each of these related values. Finally, process philosophy reminds us of the fragmentariness of all knowledge. This means that we must never become too wedded to theories or projects.

The world-view and set of values offered by process philosophy can be helpful as we think about how we will shape the future of our world. Process philosophy cannot offer a comprehensive plan for co-creating a more life-affirming future, but it can provide an overarching framework in which to situate our thinking and action. The insights and values of process philosophy can help us to locate ourselves in relation to social, economic, and political theories; to understand events, processes, and problems; and to evaluate programs for development and change. In these ways process philosophy can offer guidance as we attempt to co-create a more just and joyful world.

Social, economic, and political theories shape our understanding of the world and help us to understand how to go about changing it. Such theories are often said to be "scientific," based in "objective observation" of the way the world works. Yet all social, economic, and political theories are made from a standpoint of interdependence in a relational world and involve implicit or explicit judgments about what is real and valuable in the world. Process philosophy alerts us to this fact and provides a world-view and set of values that can help us negotiate among the claims of competing social, political, and economic theories.

For example, one of the great debates of our time has been between liberal individualist[8] or capitalist as opposed to Marxist or socialist theories of the self and society. The Cold War was fought over these competing ideologies. In the wake of the collapse of state socialism in the former Soviet Union, democracies around the world struggle over relatively more liberal or capitalist and relatively more socialist or social welfare visions of a just society. In the United States at the turn of the millennium, this debate became vicious and divisive, causing many people to become alienated from the political process. On the one side, the conservative defenders of capitalism appeal to "individual freedom" as the highest value. For them, the mature individual is imagined to be essentially separate from relationships. Each individual, it is said, has the right to determine and fulfill his or her own needs. In economic terms, "individual freedom" is understood as the freedom to create and sell products in a "free market," unrestrained by government control. Such a system, it is further argued (though this does not follow directly from the theory), will create the greatest number of jobs and therefore benefit the greatest number of people. On the other side, Marxists, socialists, and others consider "social welfare" to be the highest value. They stress the ways in which an individual is embedded in a web of social relationships that to a large extent determine her or his fate. They argue that societies must be shaped and monitored so that each individual has food, shelter, health care, and work. Marxists and socialists envision as well an equitable or more equitable distribution of wealth within societies.

What light can process philosophy shed on this debate? Process philosophy values individual freedom highly, yet it understands that individual freedom can operate only within a web of personal and social relationships. Process philosophy suggests to liberal individualists and the defenders of capitalism that individual freedom can best be exercised when the webs of personal and social relationships are structured to allow and encourage individuals to develop the capacity to make maximum use of their freedom. Where the web of relationships is not equally supportive of the freedom of all individuals, then society must be brought into greater balance. Where opportunities to exercise freedom have been limited by poverty and class division, or by racism, ethnic hatred, sexism, or heterosexism, programs must be developed structures instituted to create greater equality. Only thus will all members of a society be able optimally to express their individual freedom. To socialists, Marxists, and social welfare theorists, process philosophy recommends that social welfare programs always be directed toward the empowerment of individuals within relationships and societies. Thus, for example, programs that offer education and training and encourage individuals to take responsibility for all aspects of their lives are likely to be more successful than those that simply provide food, medicine, or jobs. State socialism becomes state tyranny when it does not recognize the importance of empowering people.

Because it is holistic, process philosophy calls attention to a dimension of social, political, and economic reality that has been overlooked by socialists and capitalists alike. This is the fact that all of our social, economic, and political arrangements exist within a web of life within which human beings are interdependent. Both socialism and capitalism failed to understand this because both accepted the (flawed) view of modern science that nature is mere matter to be controlled by human rationality. Both systems have justified the plunder of natural resources, and neither is equipped to recognize the impact of industrial and chemical pollution on the environment. Process philosophy reminds us that we must consider the environmental impact of all human projects, not only because the destruction and degradation of the environment negatively affects human life, but also because plants and animals co-created over millions of years of evolution have an intrinsic right to live.

In addition, process philosophy provides a web of interconnected values that may lead to a deeper consideration of the terms in which political, economic, and social debates are conducted. Are individual human freedom and the social welfare being understood in relation to embodiment, relationship, and enjoying life? Or are they simply being defined as the right to make money or to have certain needs met? When questions are asked in this way, the whole meaning of the debate shifts. We begin to think about our priorities in a differently. Do we enjoy life more when we have more things? Given the choice, would we rather have the money to buy things we don't really need or the time to enjoy our friends and families and the earth? How much do we value having clean air to breathe? What are our responsibilities to those with whom we share this precious time on planet earth? What will we leave to future generations?

In an analogous way, theories about the relative importance of individual freedom and social connection or relationship have shaped discussions and debates about women's liberation. The phrase "women's liberation" is rooted in the western individualist tradition, and it suggests that the goal of feminism is individual freedom for women. As we have seen in previous chapters, there is a long tradition in the West of understanding freedom and social connection as polar opposites. Traditionally, individual freedom has been a male prerogative, while women have been told that we must give up our desires for individual freedom in order to meet the needs of husbands and children. Women have also been told that we must stay in abusive relationships for the good of the family. It is not surprising, then, that women have sometimes said we want the freedom that men have always had, the freedom to put our own needs first (at least some of the time). Women do want to be free, but few of us want the freedom from relationships that is imagined in the liberal individualist view of the self. What most women want is freedom from relationships that harm the self and freedom to co-create relationships of various kinds in which our individual freedom and that of others can flourish. Process philosophy provides a perspective from which feminists can articulate our interrelated desires for freedom and more satisfying relationships.

Besides offering a way to evaluate symbol systems and theories, the world-view and values of process philosophy should be able to provide a perspective that can help us understand and respond to events and processes in our world. When two airplanes crashed into the twin towers of the World Trade Center on September 11, 2001, most Americans were deeply traumatized. Americans are used to thinking of ourselves and our government as basically good. It was hard for many of us to imagine that anyone would want to harm us. Acting out of such a self-understanding, the American president George Bush declared war on terrorism, naming Osama Bin Laden and the "forces of evil" that he had gathered together to be the enemy. Polarizing the situation into a war between good and evil, the American president reaffirmed the view that America and Americans are good. At the same time, he closed off debate both about his understanding of the problem and about whether the massive bombing of Afghanistan or the later war against Iraq that he proposed would solve it. Can process philosophy help us to think more creatively about this situation?

Earlier in this chapter, I said that it is process philosophy's view that because of the complex nature of the past and the present, it is unlikely that any problem will ever have a single cause or a single solution. A web of interrelated causes must be addressed. In chapter five, I wrote that the causes of Nazism included the fear incited by the Nazi regime, methods of discipline in early twentieth-century German families and schools, military training, long-standing currents of Christian anti-Semitism, defeat in World War I, economic deprivation, and above all the influence of Hitler. If these reflections are valid, then the first thing we can say about the events of September 11, 2001, is that there were a multiplicity of causes, and that Osama Bin Laden was only one of them. This also means that the single-factor explanation given to the American people by the American president cannot possibly be correct. What might some of the multiple causes of the September 11, 2001, terrorist attack be? I suggest that they include all of these and others: the history of western colonialism in Muslim countries; discrimination against dark-skinned people living in Europe and North America; the western tendency to view Islam as a backward religion requiring enlightenment from western sources; enormous poverty in most Muslim countries as compared with European and North American countries; a recent history of aggression by Christian or implicitly Christian nations against Islamic nations; the perception that American foreign policy favors the Jews in Israel at the expense of the Muslims in Palestine; the desire of some elements within American governments to control the world; the building of the twin towers of the World (!) Trade Center as a monument to American wealth and power; belief in infallible revelation within Islam; the rise of Islamic fundamentalism based on faith in infallible revelation; the perception that Muslim women's desires for greater freedom are inspired exclusively by the West and are wrong; the way perceptions about women are used to fuel Islamic fundamentalism; the widespread belief in almost all cultures that violence is a solution; the influence of Osama Bin Laden. If all of these causes and others contributed to the creation of the events of September 11 in New York City, then it is clear that simply eliminating Osama Bin

Laden will not solve the problem represented by the destruction of the World Trade Center and the killing of thousands of individuals in it. Process philosophy's understanding of the social construction of reality and the influence of the past on the present can help us to see that the solution to the problem of September 11 requires that all of the issues I have just mentioned and others be addressed. Most of these issues need to be addressed by Europeans and Americans. We need to gain a more balanced understanding of the history and practice of Islam. We need to educate ourselves about the history of colonialism and the way unequal economic and power relations continue to affect international politics. We need to do what we can to create a more equal balance of power in the world. We need to learn that there are non-violent methods of resolving conflict. We need to question the way we so readily assume that "God is on our side" when our nations fight wars. It is also important that liberal Muslims develop convincing alternatives to fundamentalist interpretations of Islam tied to notions of infallible revelation, and that Muslims work to create solutions to the problems their societies face, thus diminishing the appeal of fundamentalism and terror.

On the day that the World Trade towers were destroyed, my heart went out to those who had suffered and to those who died. Within hours, I began to realize that the response of many Americans was a desire for revenge. At this point my heart responded in sympathy to those who would most likely become the victims of proposed American and NATO bombing of Afghanistan. While I felt deeply for the women suffering under the restrictions of the Taliban regime, I also understood that women and children are the primary victims of war and that the Afghani land would be poisoned by a program of bombing. "Not another war," I thought, "remembering all of the wars that had been fought in my name in my lifetime." And while I despaired of being able to stop the rush to war, I sent an e-mail to everyone in my e-mail address book, urging that we not respond to violence with violence, offering instead the Nine Touchstones of Goddess ethics that I had proposed in *Rebirth of the Goddess:* "Nurture life. Walk in love and beauty. Trust the knowledge that comes through the body. Speak the truth about conflict, pain, and suffering. Take only what you need. Think about the consequences of your actions for seven generations. Approach the taking of life with great restraint. Practice great generosity. Repair the web."[9] Though my efforts and those others were taking at the same time did not stop the United States from declaring war on Afghanistan, I may have encouraged some Americans to resist the desire for revenge. My e-mail inspired a ritual held at the Montclair Women's Center in California where Linda Tillery sang and the Nine Touchstones were affirmed. Though this may seem like a trivial action with a trivial result, I believe that speaking out for peace in a climate of violence is never a trivial act. When I learned of the efforts of the Revolutionary Association of Afghan Women's efforts to educate and free women in Afghanistan, I made a contribution to the group. Later I was gratified to learn that the Global Fund for Women, to which I contribute annually, had already made a large grant to health and education programs for women and children of the Afghan Institute of Learning. In these small ways, I helped to repair the web.

An example of a problem with deeply interconnected and multiple causes is overpopulation. When the problem of overpopulation was first identified, it was considered to be a Third World problem, with large families leading to starving children. In the beginning, this was viewed as a simple problem with two simple solutions: feed the hungry and introduce birth control. Yet it was not so simple. Overpopulation is related to the lack of availability of safe and reliable forms of birth control and to lack of education about forms of birth control that are available. But birth control is not the only issue. Cultural and religious factors play a role. Many women have been told by religious authorities not to use birth control. Some feel that their husbands must make the decision about how many children they will have. Others have been told that producing many sons is the only way to increase their social prestige. Women and men may feel that they must have many children in order to insure that they will be taken care of in their old age. All of these factors need to be addressed.

In addition, the problems created by overpopulation are not limited to not having enough to eat. Overpopulation in underdeveloped countries leads to the cutting down of forests to create land for crops and wood for heating and cooking. When trees are cut down, the water table dries up and over time the land becomes desert. Many women around the world now spend much of their time walking long distances to collect small amounts of water. Thirty percent of China's land is now desert, and violent windstorms have deposited yellow dust in faraway Korea. All of these problems are real and serious.

Yet the problem of the over consumption of world resources in developed countries is equally important. Taking into account the use of the world's resources by individuals in the developed world, it could also be argued that the developed world is overpopulated. The Amazon rain forest, for example, is being cut down in part to provide grazing land for cattle that will become McDonald's hamburgers. Small families in the developed world consume far more of the world's resources than large families in the developing world. Moreover, the goal of global capitalism is to make all of us feel that we need to consume even more. This has led some to suggest that consumer capitalism, not overpopulation, is the real problem. Yet insofar as both overpopulation and overconsumption are causing great harm to people and to the environment, both must be addressed.

However, it is not easy to address the root causes of overconsumption, given that global capitalism is virtually unrestrained. Through control of advertising and the media, consumer capitalism shapes appetites and desires. Whether we are living in the developed or developing worlds, we are all being told that we must have things—more and more things—in order to be happy. In my lifetime Americans have greatly increased their appetites to consume, making the 1950s and 1960s seem innocent by comparison. When I was a child, we opened perhaps five or six presents—

among them socks and pajamas—on Christmas mornings. My brother's children each receive so many presents they get bored with opening them. I moved to Greece in the mid-1980s in part because I wanted to live a simpler life with fewer things. Since then, Greece's entry into the European Community has opened the Greek economy to foreign products, and commercial television has molded desires for more and more things. My neighbor Maro, who is younger than I am, speaks nostalgically of the time, not many years ago, when the villagers had less but enjoyed life more. Every time each of us resists the urge to buy, we are taking one small step to change our world. As societies and groups of nations, we need to find ways to limit the power of advertising funded by global capitalism to shape our desires. We need to sustain and re-create local and bioregional economies even if this means that we will not have access to every product the global economy offers us. Again, process philosophy helps us to understand that there are no single problems with single solutions in a co-created world.

Process philosophy can help us to reflect on the problems we face as we attempt to co-create a richer and more joyful world. Insofar as process philosophy leads us to identify multiple causes for the world's problems, process philosophy can also help us to recognize the need for a multiplicity of solutions. The solutions required fall roughly into three categories, although these are not mutually exclusive: hands-on-help, empowerment, and transformation of structures. The first category includes feeding those who are hungry, sheltering those who are homeless, and providing medical care to those who are sick or wounded. The second involves creating programs that can empower individuals to take control of their lives. The third involves working to change national and international laws that create or permit suffering.

"Well, yes," you might be thinking, "but if the world's problems have so many causes and require so many solutions, what can I, one small individual, do to help?" Process philosophy can help us to see that because the world is co-created, every individual act does make a difference, for better or for worse, in the web of life. Process philosophy can also help us to see that we never act alone, because we are partners with each other and with Goddess/God in the creation of the world in which we live. One thing each of us can do is to find a way to do something to improve the quality of life for human beings or other beings within our own communities. Join a Big Sister or Big Brother program. Donate time at a local homeless shelter. Organize or participate in the clean-up of a local beach or park. Cook a hot meal for an elderly neighbor once a week. Work to save a local forest. Volunteer at a rape crisis or battered women's center. Join an organization dedicated to ending war. Help elect a political candidate. Teach children to unlearn racism, sexism, or homophobia. Co-create a Passover seder, a Christian service, or a Goddess ritual that images divine power as power with, rather than power over. At the turn of the millennium, I vowed to work to save the wetlands of the island of Lesbos where I live. In 2001 I helped to collect 600 signatures on a petition to save the Gulf of Kalloni in Lesbos. An English couple who signed the petition were inspired to found the Friends of Green Lesbos, and I became a founding member of the group.[10] It doesn't really matter what cause we choose, for good causes are numerous. When we do something, we begin to give up the feeling that we are powerless and that nothing we do can make any difference. And one thing leads to another.

We can also get involved in the political process, seeking to make democracy work for more people. Suggesting this, I recognize that many of the readers of this book are alienated to a greater or lesser degree from electoral politics. In my adult life, I have never been able to feel wholehearted support for any candidate for presidential office in the United States or for any political party in my adopted Greece. On the other hand, there have been candidates in local and state elections I have greatly admired. As advertising plays a greater and greater role in election campaigns, the quality of political discussion suffers. In the United States, those with large sums of money to contribute to political campaigns, including global capitalists and conservative Christians, are playing a far larger role in political life than their numbers warrant. This suggests that in addition to voting, we need to think about how to make democracy work for us again. The roles currently played in electoral politics by advertising and campaign contributions need to be severely limited. We also need to find ways to make democracy work from the bottom up, taking into account the creative freedom of every individual, rather than from the top down. These are not easy tasks, but we must undertake them.

Another thing that each of us who has even a small amount of extra income can do is to give to non-governmental organizations that are dedicated to helping others and changing the world. Traditional religious groups often ask members to tithe or donate 10 percent of their annual income. Some years ago, two of my friends who had been giving to various causes decided to become more reflective about where their money was going and to make certain that their total giving came to at least 10 percent of their annual incomes. Their son also decided to donate 10 percent of his allowance and the money given to him by his grandparents or others. They agreed to meet periodically to talk about where each of them had decided to give their money. In this way each of them would learn more about what others were doing to save the world.[11] Talking about their giving would reinforce its importance. Giving money also helps us to understand that we are not powerless.

Inspired by my friends, I began thinking about donating 10 percent of my income to good causes. It took me some years to reach this goal, but I have now exceeded it. In the beginning, it seemed that I did not have much extra income, but the more I gave, the more I enjoyed giving, and the more I wanted to give. I have tried to give to organizations working in each of the three categories of hands on help, empowerment, and transforming structures. I have divided my giving among organizations with the aim of helping people and those with the goal of saving the environment. I am particularly interested in helping women and girls, who have been historically oppressed in most societies. The main organizations to which I give are the Nepalese Youth Opportunity Foundation, dedicated to helping children in Nepal; the Global Fund for Women, which funds projects initiated by women to empower women and girls in developing countries; and the environmental groups Greenpeace and the World Wildlife Fund, international and Greek branches.[12]

I will share part of the story of the Nepalese Youth Opportunity Foundation to show that it is possible to create and to contribute to organizations that make significant changes in the world. NYOF was created by two Americans who first visited Nepal as tourists.[13] Their sympathy for the suffering of children led to the founding of NYOF. Its initial goal was to offer hands on help to children living on the streets of Katmandu. From the beginning, NYOF has kept administrative expenses low in order to give as much of the money it raises as possible to children. Many of the street children of Katmandu are crippled or suffering from serious diseases. The NYOF has brought some of these children to the United States for operations and has sheltered and educated others in its group homes for boys and girls, J House and K House in Katmandu. Madhu, mentioned in the previous chapter, is one of the girls in the NYOF program. NYOF staff rescued Madhu's sister from a prison, where she had been living with their mother. The mother subsequently died. The father had abandoned the family. Madhu's sister cried until she convinced the NYOF staff to go out into the streets to find Madhu. At eleven, Madhu was older than the children usually taken into the group homes, and there was no bed for her. But her sister insisted that the two girls could share one bed. At first it was thought that Madhu might be too old to start school, but she followed the lessons given to the sick children in the home and within a year entered the fifth grade. She has now completed high school and is studying to be a nurse. Her goal in life is to help others. Madhu calls me her "lovely sister" because money I inherited from my parents has enabled me to provide for her support.

NYOF recently started a new program dedicated to the empowerment of pre-adolescent girls from poor villages in Nepal who are being sold into indentured servitude for the equivalent of about fifty American dollars per year. NYOF staff asked the villagers what it would take to keep the girls in the villages and send them to school. The parents said that they would need the money the girls would earn, plus school books. Further research suggested that each girl should also be given a traditional costume, since potential employers were luring girls to the cities with promises of new clothing. After NYOF enrolled a number of girls in one village in the new program, the mothers of some of the girls took staff members aside and asked them not to give money to their husbands, who, they said, would spend it on drinking and gambling. It was agreed that each family be given a piglet that it could raise on scraps and sell in the market, a kerosene lamp, and two liters of kerosene in place of the money the girls would earn. NYOF staff members found a non-local type of piglet that was said to be faster-growing than the local ones. As it turned out, the non-local piglets were not hardy in their new environment. The next year local piglets were supplied. NYOF staff also spoke with village leaders to enlist their support in the program of saving girls. After only a few years, several hundred girls had been enrolled in the program. Some villages have posted signs saying: Go away! We do not sell our girls!

The program to empower village girls can also be understood to express the values suggested by process philosophy. Sympathy for the village girls motivated NYOF staff members to create this new program. The freedom of girls is considered important. They would not be sold into servitude, and they would be educated to become self-sufficient. The NYOF staff uses a model of power with rather than power over and established relationships with the villagers. Thus they adopted the mothers' suggestion that payment be made in the form of piglets. One unexpected result of this was that the mothers also gained a greater degree of freedom, as they were the ones who fed and marketed the pigs. When the non-local piglets did not thrive, the NYOF staff were willing to reconsider their original plan. They understood the fragmentary nature of the knowledge they brought to the villages and were willing to change the program based on new information. The NYOF staff also understood the importance of taking traditional village authorities into account and gaining their cooperation where possible. A totally unexpected result at the end of the first year of the program was that a number of the Nepalese girls, who are usually shy and retiring, insisted that they be allowed to speak about the program to other girls. In its first two years of operation the program rescued 329 girls. Yet as I write this book another 1,500 girls are being sold in the single valley in which the program was begun. This year I sent part of the money that my father gave me for Christmas to the NYOF program for saving girls. I have enjoyed the thought that I have helped to save girls in Nepal far more than I would have enjoyed spending the money on myself.

For process philosophy the reason for hope is the co-creative process of life itself, a co-creative process that is supported, sustained, and shared by Goddess/God. For process philosophy the universe is co-created by individuals working together with each other and with Goddess/God. Because individuals really are free, our contributions can make a difference to the world. Because individuals really are free, the future of the world cannot be known, not even by Goddess/God. Human beings have the power to help to co-create a more just and joyful world for each other and for all beings in the web of life. Goddess/God supports and sustains this effort. Human beings also have the power to destroy the web of interconnection that allows the flourishing of human and other forms of on planet earth. Goddess/God cannot guarantee that life abundant will continue on our planet. We hold the future of life on earth in our hands. May we choose life.

If the world really is co-created by all of the individuals in the world, then why have philosophers and theologians so often said that God knows the future of the world? Could it be that the theological mistake of asserting divine omniscience, like the others to which it is so closely connected, stems from a rejection of finitude and interdependence that begins with a rejection of the female body?

Chapter Eight

Restoring the Body and the World

She changes everything She touches and everything She touches changes. The world is Her body. The world is in Her and She is in the world. She surrounds us like the air we breathe. She is as close to us as our own breath. She is energy, movement, life, and change. She is the ground of freedom, creativity, sympathy, understanding, and love. In Her we live, and move, and co-create our being. She is always there for each and every one of us, particles of atoms, cells, animals, and human animals. We are precious in Her sight. She understands and remembers us with unending sympathy. She inspires us to live creatively, joyfully, and in harmony with others in the web of life. Yet choice is ours. The world that is Her body is co-created. The choices of every individual particle of an atom, every individual cell, every individual animal, every individual human animal play a part. The adventure of life on planet earth and in the universe as a whole will be enhanced or diminished by the choices we make. She hears the cries of the world, sharing our sorrows with infinite compassion. In a still, small voice, She whispers the desire of Her heart: Life is meant to be enjoyed. She sets before us life and death. We can choose life. Change is. Touch is. Everything we touch can change.[1]

This depiction of process philosophy's Goddess/God as She Who Changes contrasts sharply with the picture of God as an Old White Man with a long flowing beard discussed in chapter one. Process philosophy's understanding of the divine in the world challenges the assumptions of classical theism that are expressed in the traditional image of God. Whereas classical theism begins with a denial of the value of changing life, process philosophy affirms that change is inherent in every individual life in the universe at every moment and that change is the nature of the evolutionary process of life as a whole. Human beings were not created half way between animals and angels, as classical theism asserts. We evolved through a process of change within the web of life, sharing the capacity to feel and to feel the feelings of others and to exercise creative freedom with all other individuals. Goddess/God changes with the experiences of every individual in the changing world that is the divine body, while remaining unchangeable only in one respect: Goddess/God will always and everywhere relate to the world with creativity and love. Classical theism considered relationship to be a limitation and envisioned living alone or apart from others to be the most perfect state for both divinity and humanity. Process philosophy affirms touch or relationship as fundamental for all life, including divine life. In classical theism, divine power is unlimited. An omnipotent God is in control of the world; everything—even what appears to be evil—happens according to divine will and purpose. Process philosophy says that the power of Goddess/God is power with, not power over. The assertion of God's absolute and unlimited power creates the problem of evil: How could a good and loving God allow so much suffering to exist? Process philosophy answers that Goddess/God did not create suffering. The world is co-created by every individual in it. Death is part of life, but much of what we know as suffering is created by human beings. Goddess/God is with us in our suffering and inspires our efforts to lessen or transform it. Classical theism states that the goal of human life is to rise above the changing body and to share in the immortal life of God.

Process philosophy asks us to enjoy finite and changing life that ends in death. The omnipotent God of classical theism is said to be the author of infallible revelation, given in the form of texts, teachings, traditions, or inspired individuals. Process philosophy counters that all human knowing is embodied and embedded in the world—and that it will always be fragmentary and in process. Classical theism asserts that an omniscient God already knows the fate of the universe. Process philosophy asks us to consider that the future of the universe is unknown even to Goddess/God. Though shaped by the past, the future will be created by the choices of a myriad of individual wills. This means that the outcome of human moral efforts to save or improve the world cannot be assured. The reason for hope is the open-ended creative process of life itself, which is supported and sustained by Goddess/God.

Process philosophy shares with feminist theology and thealogy a common interest in restoring the body and the world body, disparaged and denied in classical theism. What process philosophy has frequently failed to recognize is that restoring the body and the world body has enormous consequences for women. A feminist process paradigm will make feminist insight an integral part of process thinking. A feminist process paradigm will also ensure that process philosophers understand the body, the world body, and the divine body in physical terms and not simply as metaphysical concepts.

Feminist theologians have long recognized that women have been viewed as secondary or subordinate in dualistic anti-body traditions that follow Plato in making a sharp distinction between God and the unchanging soul on the one hand, and the changing body and nature on the other. In dualistic philosophies created by men, the rational soul of man is associated with the unchanging immortal realm of (a male) God, while woman is identified with the body, nature, and death. Feminists have called this way of thinking hierarchical dualism because one set of qualities—the unchanging, the rational, the soul, the male—is valued more highly than the other—the changing, the natural, the body, the female. In such traditions God must be imaged as male because maleness is associated with the unchangeable realm of soul and spirit. God cannot be imaged as female because femaleness is associated with the changing body, nature, and death.[2]

In light of this analysis, I asked a simple question at the end of the previous chapters of this book: Is the source of the theological mistakes of classical theism a rejection of embodied life that begins with rejection of the female body? In other words, are the six theological mistakes embedded in a way of thinking that is inherently anti-female? I suggest that the answer to this question is yes. The six theological mistakes are based in denial of the changing body and the changing world, which is rooted in a way of thinking that is inherently anti-female. "But," it might be objected, "anti-body thinking need not be anti-female. It only became so because men projected the body they despised onto women. After all," it might be added, "women are just as capable of rejecting the body as men. Look at the female saints who starved themselves to death. If women had created dualistic traditions, they might have been anti-male."[3] It is true that women have been enthusiastic supporters of anti-body traditions in the west and in the east. Still, it is not sheer coincidence that dualistic traditions have a particular antipathy toward the female body. Dualistic traditions reject not only the body but the whole physical world of which the body is a part. Inevitably they take a negative view of physical birth through a female body.

Indeed, many dualistic traditions contrast (inferior) physical birth to (superior) spiritual birth or "re-birth." Socrates called himself a midwife of the soul. Circumcision rites claim male blood as more holy than the blood of birth. Baptism names water consecrated by men more powerful than the waters of birth. Enlightenment has been understood as release from the realm of birth and death. Dualistic anti-body philosophies originated in cultures in which more ancient traditions made a positive association between the body of the mother and the creative powers of Mother Earth or Goddess.[4] Dualistic traditions reject not only the female body through which we enter the physical world but the physical world itself, the earth that was known as the body of Goddess. If this is true, then we must recognize that dualistic traditions are not only inherently anti-female, but also matricidal.

But is calling dualistic traditions matricidal going to far? It could be argued that denying the body that gives us birth is not matricide or mother-murder but technically speaking mother-denial. However, the tradition of Platonic dualism arose in the ambience of the repeated performance of the *Orestia* of Aeschylus. In this trilogy, Orestes avenges the death of his father by killing his mother. At his trial the Goddess Athena justifies her vote in favor of the son who murdered his mother with the words, "No mother ever gave me birth."[5] There are good reasons for considering that rejection of the body of the mother is at heart matricide. The mother who is murdered is not only the human mother, but also the physical world, the mother of all.

In challenging dualistic thinking, process philosophy pulled the rug out from under anti-female and matricidal habits of thought. Yet if process philosophy is implicitly feminist, then why did Whitehead and Hartshorne not state explicitly that process philosophy is a feminist philosophy? I think it is fair to say that this thought never occurred to either of them. Commenting on the history of science, Whitehead said, "There are some fundamental assumptions which adherents of all the variant systems within the epoch unconsciously presuppose. Such assumptions appear so obvious that people do not know what they are assuming because no other way of putting things has ever occurred to them."[6] I suggest that the idea that whole traditions of philosophical thinking had for centuries been systematically anti-female is an idea that simply did not occur to either Whitehead or Hartshorne.

What are we to make of this rather serious "oversight"[7] of these two great thinkers? The conclusion I would draw is that they were men of their own times, not ours. Whitehead and Hartshorne knew that some philosophers were male chauvinists,[8] but they assumed that philosophical systems were gender neutral because no other way of putting things ever occurred to them. That another way of putting things has occurred to us is testimony to the revolutionary power of the feminist movement of the last three decades of the twentieth century. Feminism has taught us to scrutinize all theories for explicit and implicit gender bias. A feminist process paradigm can help us to see that the theological mistakes of classical theism are disproportionately harmful to women because they are based in rejection of the female body through which we are born into the physical world

If the theological mistakes of classical theism—conceiving of divine power as unchanging and unsympathetic, omnipotent and omniscient, and the related views of infallible revelation and immortality—are not only inherently anti-female but also rooted in matricide, then spiritual feminists ought to avoid them at all costs. A feminist process paradigm can help us to recognize the importance of the alternatives to classical theism offered by process philosophy for our feminist efforts. In fact, many feminist theologians and thealogians, myself included, have been moving in the direction of affirming the process values of change, embodiment, relationship, sympathy (understanding, love, or compassion), creative freedom, enjoyment, and power with rather than power over. A feminist process paradigm can help us to recognize that these values are interconnected and thus require each other. It can also provide a fresh perspective from which to think about issues that have become controversial within feminist theologies and thealogies.

A feminist process paradigm can shed light on conflicts that have arisen in relation to attempts to reclaim the female body, the earth body, femaleness, and the feminine in feminist spirituality. Early on, spiritual feminists understood that affirming the changing body and especially the changing female body was a priority. At the first gathering of women theologians at Grailville in 1972, a widely reprinted retelling of the story of Adam and Eve entitled "The Coming of Lilith" challenged the symbolic association of the female body with sin and evil.[9] At about the same time, a group of women created a summer solstice ritual in which they formed a symbolic birth canal in order to birth each other into a circle of women. They raised power by placing their hands on each other's bellies. They marked each other's faces with menstrual blood saying: "This is the blood that promises renewal. This is the blood that promises sustenance. This is the blood that promises life."[10] In the ensuing years, Audre Lorde's groundbreaking essay "Uses of the Erotic: The Erotic as Power" was adopted as a kind of manifesto by Jewish, Christian, and Goddess feminists because it gave us permission to trust the feelings—including sexual feelings—of joy and pleasure that we experience in our bodies.[11] Women also began to create a wide variety of rituals to celebrate menarche, birth, and menopause. The popularity of images of the naked female body of the Goddess in feminist art and ritual is testimony to women's hunger for symbols that express the creative and sacred powers of the female body.[12] This theme was picked up in the controversial Christian feminist prayer to Sophia that reimagines *"women in your image: With warm nectar between our thighs."*[13] Though yoga is often said to be a method that leads to the transcendence of the body, feminist teachers and practitioners are more likely to understand yoga as a way of finding the spirit in the body and in the body as female.[14] Similarly feminists tend to view meditation not as a way to escape the world of "birth-and-death," but to learn to live more comfortably within it.[15] Like yoga, meditation can help us to become aware of the connections between body, mind, and spirit.

Much feminist work in religion has focused on relationship and interconnection, alleging that women's bodily experiences attune women to relationships, that women's way of knowing is through relationships, or that women have a special ability to sense the human connection to the earth or and all of life. Many women have been drawn to Jungian thinking because it values aspects of life that have been traditionally been called female or feminine, including the body, the unconscious, intuition, and relationship.[16] Some Christian feminists have begun to develop a more relational view of the world, or what they call a theology of "mutuality."[17]

Many of them, along with Jewish and other feminist thinkers, have been inspired by Martin Buber's depiction of I-thou relationships with other people, with nature, and with God. Goddess feminists, ecofeminists, Asian feminists, and western women inspired by non-western religions speak of the web of relationships in which we are enmeshed as "the web of life" and they urge greater concern for nonhuman life.[18]

At the same time the meaning of the body, the female body, relationship, and interconnection have become highly contested questions within feminist thinking in religion. While many women have felt affirmed by images, rituals, and theories that celebrate the female body, relationship, and interconnection, others have found them limiting. Critics ask whether some feminists end up affirming conventional gender dichotomies, even if we do not intend to do so. Do rituals that celebrate menstruation, birth, and menopause reinforce the idea that woman's primary role is motherhood and perhaps even that her place is in the home? Do images of the naked female body of the Goddess simply continue a long tradition of associating women with the body and nature, but not with the mind and freedom? What about the male body? What about women who do not have children? Women whose sexuality is directed to other women? Women who do not have partners? Women in abusive relationships?

Some academic feminists use the term "essentialism" (as in women are "essentially" embodied and relational) to characterize feminist standpoints that they feel lock women into one set of possible roles and behaviors while excluding others.[19] They point out that theories that assert a dichotomy between masculine and feminine or male and female characteristics or ways of being limit the possibilities of both women and men. Those who are uncomfortable with the emphasis on the female body, relationship, and interconnection in the thinking of other feminists have legitimate concerns. It is important to value the female mind as much as the female body. It is important to affirm that women are free and creative individuals. It is important to recognize that men as well as women are embodied. It is important to remember that relationships come in many varieties and that they can harm as well as nurture and heal. It is important that feminists not create new stereotypes that limit women's creativity and freedom.

A feminist process paradigm offers a perspective in which the concerns of both sides in what has become a sometimes acrimonious debate can be taken into account. In process philosophy, the importance of body, relationship, and connection to nature is affirmed, as in many feminist theologies and thealogies. Process philosophy also clearly understands that men as well as women are embodied, related, and interconnected in the web of life. The male body as well as the female body changes and dies. Men as well as women are internally related to others. Men as well as women are part of nature. Process philosophy offers a holistic understanding of the world in which change and embodiment and interdependence are affirmed for all individuals—women, men, animals, cells, the smallest particles of an atom, and Goddess/God. Process philosophy also affirms that all individuals are free and creative. In such a world women and men can express creative freedom within our bodies, relationships, and connections to changing life. Process philosophy legitimates feminist re-imaginings of divine power through the lens of the female body and its connections to nature. Yet process philosophy reminds us that all bodies—male as well as female, nonhuman as well as human—are part of the divine body and thus can function as images of divinity.

There is no doubt that if compared to traditional stereotypes of masculinity and femininity, process philosophy's understanding of Goddess/God is more feminine than masculine. The divine sympathy, as Hartshorne said, has more in common with traditional understandings of a mother's love for her children than it does with traditional notions of masculine power, including traditional notions of fatherhood. Like many feminist theologies and thealogies, process philosophy stresses what has been called the divine immanence—change, embodiment, relatedness, and interconnection.[20]

Yet process philosophy also says that there also are ways in which Goddess/God is transcendent. It can thus help us to affirm the traditionally feminine attributes of embodiment and relatedness while not losing sight of human, nonhuman, and divine freedom and creativity, which have usually been thought of as masculine and transcendent.

While asserting that all individuals are embodied, related, and interconnected, process philosophy views all individuals as free and creative—from the smallest particle of an atom to Goddess/God. Thus for process philosophy there is no question of choosing mind over body or body over mind. Mind and body are one continuum. It is as true to say that all individuals are free and creative as it is to say that all individuals are embodied and related to others. In process philosophy all free and creative individuals are embodied, and all embodied individuals are related to other free and creative individuals. To imagine that freedom is somehow antithetical to relationship and embodiment (as many philosophies and many individuals, including some feminists, have done) is quite simply a mistake in thought. This does not mean that embodiment never limits the choices we have. All choices are limited choices. To wish it to be otherwise is also a mistake. Nor does it mean that individuals never experience conflicts between the responsibilities inherent in relationships and individual needs for free and creative expression. But there is no essential or necessary conflict between embodiment and relationship and interconnection on the one side and creative freedom on the other. Nor is there any need to state that women are more embodied, relational, and interconnected than men, for these qualities define all individuals. Process philosophy suggests that rather than arguing about who is right and who is wrong, feminists might join together in imagining a world in which embodiment, relationship, and creative freedom are understood to be components of all life in the universe.[21]

A feminist process paradigm sharply poses the question of the meaning of death in theologies and thealogies that affirm life in the body. Valerie Saiving's groundbreaking essay "The Human Situation: A Feminine View" launched the current wave of the feminist theology movement.[22] In a less well-known essay on feminism and process philosophy, Saiving prophesied: "The most basic assumption we have inherited from patriarchal culture, the one which feminists may find the most difficult to overcome, is that the enduring self is the true locus of value, and that the death of that self is the greatest adversary."[23] Saiving saw the dualism of life and death as the source of all the other dualisms, including those between masculinity and femininity, rationality and emotionality, freedom and relationship. Though Saiving does not say so explicitly, this can be true only if the fear of death is a consequence of a rejection of life in the body, life that comes through the body of the mother. Saiving further suggests, following Whitehead, that the fear of death is related to the idea that the "enduring self" is the greatest value in life. If it is not, then what is? The answer is simple: the process of life itself. Valerie Saiving had already begun to suffer from disabling arthritis when she asked us if we could learn to include death within the process of life. Two decades later and apparently unaware of the way in which Saiving had anticipated her insight, Grace M. Jantzen suggested that western (male) philosophers and theologians have been obsessed with death and immortality because they ignored or despised "natality," birth and life.[24]

Perhaps because feminists have intuitively understood that concern with death and immortality begins with a rejection of birth, consideration of life after death has not been a focus in feminist work in religion. I have long suggested that Goddess religion teaches reverence for life that enables us to affirm finitude and death.[25] Rosemary Radford Ruether, writing from the perspective of an ecofeminist and cosmic understanding of Christianity states that that this life rather than life after death should concern us.[26] Far more than Christianity, modern Judaism, including Jewish feminism, has focused on life in the body and this world rather than on the hope for life after death. However, as I noted earlier, some Goddess feminists and followers of non-western religions have found eastern theories of reincarnation attractive. Nor have all Christian and Jewish feminists concluded that the hope for life after death is inconsistent with affirming the body. Process philosophy asks us to consider whether belief in immortality or reincarnation is a legacy of dualistic thinking, east and west. Feminism sharpens the questions: Is the feminist intuition that life in the body and the world is to be affirmed and enjoyed compatible with belief in life after death? Do theories of re-birth and re-incarnation suggest that birth through a female body into a world that includes death simply isn't good enough? As the first and second generations of feminist theologians reach an age at which death is more present in our lives, perhaps we will reflect together on these questions.

Another area of controversy within feminist theologies and thealogies concerns the meaning of new symbols of Goddess/God as She. While spiritual feminists share a critique of the distant and judgmental God-out-there of traditional piety, we are not always clear about how to define the relation of Goddess/God to the world. Is God-She or Goddess immanent or transcendent? Should we call ourselves pantheists? Mystics? Polytheists? Monotheists? What do we mean by these terms? A feminist process paradigm can help us think about these questions as well. I believe that process philosophy's panentheism affirms the best insights of monotheism, mystical pantheism, and polytheism while avoiding the mistakes associated with traditional dualistic theism. Panentheism means that that the world is "in" Goddess/God.[27] In process thought, the world is "in" Goddess/God because Goddess/God sympathetically feels the feelings of every creature. Insofar as the world is also understood to be the body of Goddess/God,[28] panentheism can also be said to mean that Goddess/God is "in" the world, is felt by and included in every creature. Panentheism differs from classical theism in not radically separating God from the world. Yet panentheism shares with monotheism "the intuition of unity," the sense that the divine power is a unifying principle in the world.[29] The divine power keeps the creative process of the universe from deteriorating into chaos. Panentheism shares with biblical and other religions the understanding that the divine power can be related to as Thou, that divine power is personal and cares about individuals and the world. Yet more clearly than biblical religion, panentheism asserts that divine power is always the power of persuasion, not coercion, power with, not power over. Because it understands the world to be the body of Goddess/God, panentheism agrees with mystical pantheism that the world is sacred. Though not asserting a multiplicity of divine powers, panentheism, like polytheism, is open to a wide range of images and symbols for divine power, as many as the individuals that make up the divine body.

"Panentheism" is not a word that is likely to slip easily into everyday language. Still, I suggest that feminists who have rejected the God-out-there of dualistic theism may find that panentheism expresses our new understanding of the divine in the world more clearly than theism, monotheism, pantheism, or polytheism. Theism is associated with dualism, radical separation, and power over. Monotheism carries the baggage of religious intolerance. Pantheism has difficulty asserting that individuals other than Goddess/God really exist. Polytheism denies that there is a unity underlying multiplicity. Panentheism affirms divine presence in a co-created world.

Panentheism also offers a new way of thinking about the question of how individuals are related to the whole, the question called "the problem of the one and the many." From traditional perspectives there are two options: either individuals are real, or all is one. Both of these options are problematic. To say that individuals are real takes account of our common sense understanding of individuality, difference, and freedom. Yet from most traditional perspectives, it leaves individuals separate from each other and from the divine ground and power. Against this view, monists have asserted that contrary to ordinary perception, all is one. Yet monism cannot explain individuality, difference, and creative freedom. Because it offers an understanding of the relationship of all individuals to each other and to the divine ground that does not obscure individuality, difference, and creative freedom, process philosophy offers new way of thinking about an age-old question.

Panentheism is compatible with the intuitions of Goddess feminists and ecofeminists that the earth and its processes of birth, death, and regeneration are sacred. Similar views are found in the symbols of indigenous religions, for example in the images of Changing Woman of the Navajos and Oshun the Goddess of flowing waters of African and African American religions. Acceptance or affirmation of change in non-western religions such as Taosim or Buddhism is one reason that many feminists have turned to them. Christian ecofeminists imagine the world as the body of God.[30] Some Jewish feminists understand Shekhina as "She Who Dwells Within" (the changing world).[31] Yet feminists who affirm Goddess/God's relation to the changing world are often accused of romantically clinging to pre-modern ways of viewing the world that cannot be reconciled with science.

Process philosophy can help us to affirm ancient visions of the changing earth as sacred while expanding traditional understandings to include evolution. Ancient and traditional peoples imagined the starry heavens as a backdrop to life on earth. Yet we know that the earth is but one of the planets that circle our sun, which is one of millions of stars in the universe. It is unlikely that life occurs only on planet earth. While it was appropriate for ancient peoples to speak of earth as Goddess, it may be more appropriate for us to understand the world or the universe as the body of Goddess/God. A feminist process paradigm supports the feminist intuition that the earth body is sacred, while helping us to clarify our thinking and expand our vision. It also makes it clear that the desire to save the earth and its creatures is not misplaced romanticism.[32]

Panentheism can also help us to think about questions raised by the feminist appropriation of a way of thinking about God called the "via negativa," or negative way.[33] In order to justify criticism of biblical images of God as Lord, King, and Father, spiritual feminists have frequently appealed to the traditional argument that all images of God are limited because they are human creations. "Biblical images," we have said, "are the creations of men. As such they have no eternal validity." Mary Daly and more recently Elizabeth Johnson have asserted that the negative way of Roman Catholic theology can be a helpful tool for feminist theology.[34] The negative way states that because God is more than any human conception, it is easier to say what God is "not" than to say what God "is." For example, we may say "a mighty fortress is our God, " but we know that God is "not" really a fortress. If the negative way is true, then God is also "not" a Father, "not" a Lord, and "not" a King, even though traditional religions may claim divine sanction for these images. If no images of God are literally true, then female imagery for God is just as appropriate as male imagery. The negative way justifies Elizabeth Johnson's use of the radically new name "She Who Is" for the divine power. Johnson cleverly turns tradition against itself. Protestant theologian Sallie McFague uses a similar logic when she says that all of our language about divine reality is "metaphor" because God is in some sense wholly other.[35] Laurel Schneider claims that all feminist theology must assume God's radical otherness, for only from this standpoint can feminists relativize biblical language and imagery.[36]

Process philosophy, as we have seen, asserts that all human knowing is fragmentary. But this is not the same as saying that we can know "nothing" about Goddess/God. The problem with the negative way is that it is rooted in the dualistic understanding that there is an absolute discontinuity or impassible gulf between divinity and humanity. In other words, the negative way is rooted in the notion of radical divine transcendence of the world. Yet this notion is rooted in the first theological mistake, the view that God does not participate in the changing world. From the standpoint of the negative way, which relativizes everything we say about Goddess/God, it is not possible to say that the divine power is love rather than hate. Theologies built on the negative way often appeal to revelation to bridge the gulf between divinity and humanity. If the revelation is associated with the Bible or any traditional religion, then all of the problems associated with patriarchal traditions are reintroduced. If, on the other hand, contemporary experience is the source of new revelation, the negative way can tell us only that all images of divinity are relative. This protects against the danger of imagining that any particular understanding of divinity is absolute. Beyond that it does not provide any concrete help in evaluating religious symbols.

In contrast to theologies of the negative way, process philosophy can construct positive alternatives to traditional understandings and imagery, because it does not begin with the notion that the difference between Goddess/God and the world is absolute and unbridgeable. Panentheism asserts that the divine power is in some sense immanent in the world. As notions of divine transcendence created the negative way, so notions of divine immanence gives rise to a "via positiva" or positive way, asserting the likeness of divine power to the powers of other individuals in the world. In process philosophy, Goddess/God is in significant ways "like" other individuals. All individuals from the smallest atom to Goddess/God share embodiment, relationship, and creative freedom, though in vastly differing degrees. Because of this, we can know "something" about Goddess/God in and through knowing the world. Our understanding is always fragmentary, but it can provide a more sophisticated tool for evaluating theologies and religious symbolism than the negative way. A feminist process paradigm suggests that the negative way will not serve feminist interests in the long run because it is based in the first theological mistake, a way of thinking that begins with denying the body.

Panentheism also provides a way beyond a related problem known as the "anthropomorphic fallacy." While spiritual feminists reject the God-out-there because we sense that divine power is intimately related to the world, we have also been taught to be wary of making the error of anthropomorphism or attributing "human" qualities to God. Though we may invoke Goddess, Sophia, or Shekhina in our prayers, meditations, or rituals, we may shy away from asserting that we think of the divine power as a "person" who "cares" about the world. We have been told that it is "childish" or "egocentric" or "irrational" to imagine that the divine power cares about our individual lives. And we may worry that if we think of divine power as personal, we will end up re-imagining God as a dominating other. What is not recognized is that the critique of anthropomorphism stems from notions of God's absolute difference from and transcendence of the world.

As we have seen, process philosophy asserts that in many significant ways divine power is "like" the power of other individuals in the world. From within the system of process philosophy, the divine sympathy is the most accurate way of describing divine power. This means that though Goddess/God is not "a person exactly like ourselves," Goddess/God is "personal" if by that we mean Goddess/God is "the eternal Thou," capable of feeling and of feeling the feelings of others in a profound way. For process philosophy, as we have seen, all individuals are capable of feeling and feeling the feelings of others, though in vastly differing degrees. To affirm this capacity of atoms and cells yet to deny it to Goddess/God does not make any sense from a process point of view.[37] This does not prove that Goddess/God exists, but it does show us that it is not an anthropomorphic fallacy to speak of the divine power (if we are going to speak of it at all) as caring and loving and therefore as personal.

For me, process philosophy's understanding of the divine sympathy and therefore of the personal nature of Goddess/God is a strength, not a weakness. The notion of the divine sympathy makes sense of the widespread use of personal imagery for divine power in most of the world's religions as well as in feminist spirituality. The divine sympathy is a way of understanding the experience that many of us have had that divine power is present in our lives, sharing our joy and our suffering. For me the divine sympathy is a way of naming the divine presence that

I experienced dramatically in times of crisis and that I also experience daily in less dramatic ways. I find personal imagery for Goddess/God necessary and appropriate, because I experience Goddess/God as caring about my life and the lives of all other individuals in the world. It is sometimes said—in the Goddess movement as well as in Christian and Jewish contexts—that we use anthropomorphic imagery because it affects us deeply, even though we don't really believe that the divine power is personal. But if this really is our view, then I think it would be more appropriate to dispense with personal imagery, as Marcia Falk does in her later prayers.[38]

This discussion of the negative way can help us to situate feminist theologies and thealogies in relation to deconstructionism. Insofar as it assumes that all human knowledge is fragmentary, process philosophy agrees with deconstructionism that issues of perspective and power are involved in all constructions of truth. Like deconstructionism, process philosophy states that we cannot know the truth with absolute certainty or express it in a global or universal fashion. On the other hand, when radical deconstructionism says that we can therefore know "nothing" about reality, it bears an uncanny resemblance to the negative way in which it may be rooted. In contrast to radical deconstructionism, process philosophy asserts that fragmentary knowing is possible and that we can talk about the greater and lesser adequacy of ideas and frameworks for understanding embodied embedded life. Process philosophy also suggests that not all personal and social relationships are based in the will-to-power or power over, as deconstructionism asserts. For process philosophy, power with is possible. At the same time, deconstructionism, like Marxism and Freudianism, can help us to understand and explicate the ways that power over functions in personal relationships and contemporary socially constructed worlds.

A feminist process paradigm identifies trust in infallible revelation of any form as a theological mistake rooted in a denial of finitude. A critique of authorities—traditional and nontraditional—is implicit in all feminist thinking in religion. There would be no feminist spiritual movement if feminists accepted the authority of traditional religions or traditional spiritual leaders. There is a growing recognition among spiritual feminists that though we may choose to situate ourselves within traditions, we can appeal to no authority outside of ourselves and our communities as we seek to transform, re-imagine, and re-create new understandings of the divine in the world. For Roman Catholic Elisabeth Schüssler Fiorenza, not the Bible but the community she calls the "*ekklesia* of wo/men," the church of women and men seeking equality and liberation, is authoritative.[39] Jewish theologian Judith Plaskow argues that feminists and other liberal Jews should acknowledge that they are choosing, among the various strands of Jewish tradition, which ones to emphasize in creating a renewed Judaism.[40] Increasingly, participants in the Goddess movement are realizing that we are creating new religious traditions, selectively drawing on the past.[41] As western feminists become more familiar with non-western traditions, they recognize that they must exercise creative freedom in relation to them, choosing the elements of traditions that speak to them, while rejecting or reinterpreting those that are not fundamentally life-affirming.[42] If this is so, then we are all engaged in creating new syntheses within religious traditions. Yet the habits of trusting authorities and seeking certainty have been strongly encouraged by our cultures. It is easy for these habits to slip back into our thinking. A feminist process paradigm can help us to understand that if we affirm the body we must also accept the fragmentariness and uncertainty of all human knowledge. At the same time, this frees us to re-imagine the divine and the world from the perspectives of our own experiences.

Having discussed some of the ways that a feminist process paradigm might illuminate feminist work in religion, I would now like to consider several questions that feminists have raised about the adequacy of process philosophy's understanding of the divine in the world. Each of these questions was posed to me by other feminists while I was working on this book. A few of these questions arise from a misunderstanding of process philosophy. Some of them suggest ways that feminism can make a contribution to process philosophy, challenging and deepening its understanding of important issues, while others call attention to ways in which process philosophy challenges some aspects of feminist theologies and thealogies.

Some feminists have asked if the process view of the divine sympathy creates a picture of divine power that is "too good to be true." This is an interesting reaction. From most traditional points of view, the Goddess/God of process philosophy would be dismissed as not powerful enough to merit the name "Goddess/God." It is the omnipotent God of traditional theology and piety who will make all things well that is usually said to be a wish fulfillment. Do some feminists find the process view of divine power "too good to be true" because we have been taught to think of God in terms of power and control, not love and understanding? Do we find it difficult to accept that Goddess/God wants us to enjoy our lives, having been taught too well that to suffer is divine?

Or is the objection rather that process philosophy idealizes divine power by associating it more with good than with evil? Some feminists have been attracted to images of Hindu Goddesses who create and destroy precisely because they do not idealize divine power. Isn't Goddess/God equally as present in earthquakes and small pox as in sunny skies and healthy bodies? In rage as well as in love? In war as well as in peace? Do we really want to sanitize divine power by imagining it as Good? Especially when there is so much evil in our world? These are important questions and I doubt that all feminists will agree on the answers to them. Yet I find process philosophy's answer compelling. Insofar as the world is the body of Goddess/God, both good and evil are included within Goddess/God's body. Goddess/God is "with us" in sickness and in health, in suffering as well as in joy. However, though all is included in the divine body, Goddess/God is not equally present in good and evil, love and hate, creation and destruction. As the soul of the world body, as a personal presence in the world,

Goddess/God is always on the side of good, which is to say on the side of understanding, compassion, and love. Goddess/God is always trying to help us to co-create the conditions in which the greatest number of individuals can enjoy life. Some feminists may find that this is to limit Goddess/God. They prefer to say that divine power is unlimited in the sense of being "All" in an unqualified way, equally good and evil. Yet, as discussed earlier, the question arises as to whether a Goddess/God that is equally inclusive of good and evil is indifferent to the suffering in the world. If so, is such a divine power one with whom we would wish to enter into relationship? Is a divine power that can make no distinction between loving a child and raping it, or between Auschwitz and a loving home, worthy of our worship? Hartshorne answered this question when he said, "Either God really does love all beings . . . or religion seems a vast fraud."[43] I agree with Hartshorne that a divine power that is not compassionate and loving is not one I would want to worship. Still, I find it unlikely that either intellectual or moral arguments that Goddess/God is love, compassion, and understanding will convince anyone who does not already have an experience or intuition of the power of love divine, all loves excelling.

On the other hand, some Christian feminists have no quarrel with notions of the divine sympathy and love, but find a feminist process paradigm unnecessary because these values can be found in Christian theology. There is no question that the notion that Goddess/God is love can be found in Christianity. Whitehead, Hartshorne, and I, along with many others, learned about the love of God in Christian churches. But Whitehead and Hartshorne did not understand the views of the world they proposed to be narrowly or exclusively Christian. Hartshorne said that Jewish theologian Abraham Joshua Heschel's understanding of the divine in the world was remarkably close to the process view, though worked out independently.[44] There are many connections between Martin Buber's world-view and that of process philosophy. Both Whitehead and Hartshorne were aware that Buddhism anticipated the view of the self they presented. Hartshorne felt that there are strands within Hinduism that are very similar to process thinking.[45] There are connections between the Tao and process. Many indigenous traditions share with process philosophy the understanding that the world is alive and that the divine power is present in it. And though the notion of God as love has been identified with certain forms of Christianity, this idea is not uniquely Christian, nor has it always been central in Christianity. In Christianity the notion that God is love is bound up with many other ideas—including male dominant language and imagery, the six theological mistakes of dualistic traditions, and the problematic notion that the love of God is somehow expressed through the sacrifice of his beloved son. Process philosophy can acknowledge its debt to Christianity in affirming that Goddess/God is love while clearly distinguishing its vision from biblical language and imagery, the errors of classical theology, and the emphasis on sacrifice in (western) Christianity.[46]

Other feminists fear that the idea that Goddess/God is love is limiting in another way. They ask whether the idea of divine love calls to mind the moralistic and repressed, colorless, asexual, always sweet and clean, never angry or dirty, self-sacrificing model of Christian womanhood handed down from the nineteenth century. Many of us have known mothers and grandmothers, schoolteachers and Sunday school teachers who exude "Christian love," yet, just underneath it, are seething with repressed anger. Inevitably their love feels like control. Is the disembodied, repressed, and self-sacrificing Christian love that so many have experienced what process philosophy means when it says that Goddess/God is love? This question is easily answered from a process perspective. Process philosophy says that life is to be enjoyed in the body and in relationship. Those who love life share this love with others. Loving one's neighbor as oneself is part of relational life. But self-sacrifice is neither the goal nor the meaning of life.

The image of repressed and self-sacrificing love raises another and more critical feminist question about process philosophy's understanding of Goddess/God as love. Is expressing anger compatible with love?[47] Whitehead and Hartshorne did not address this question directly. In the liberal Protestant Christianity of their upbringing, biblical depictions of the divine anger were assumed to reflect lower stages in the development of religious consciousness. Hartshorne wrote about the joys of a sometimes stormy relationship with his wife Dorothy, but he never addressed the relation of anger and love from a philosophical point of view. It goes without saying that neither Whitehead nor Hartshorne understood himself to be part of an oppressed group seeking liberation.

In contrast, feminists and others know the healing power of anger when it allows us to react against injustice—both personally and politically. "You can't do that any more—not to me and not to us!" we have said. "No! Never again!" For many of us, expressing anger has been a powerful stimulus to personal and political transformation. Feminist ethicist Beverly Harrison challenged narrow understandings of Christian love when she wrote that expressing anger can be a work of love, offering others a chance to learn and encoding the hope that change is possible.[48] In my life, expressing anger at those who uphold structures of injustice has helped to create change. In my spiritual journey, expressing anger at both God and Goddess has played a pivotal role, as I have discussed earlier. Over the years I have also learned to express anger in my personal life rather than letting it simmer into bitterness. Yet women—and I am one of them—know the destructive power of violent anger turned against us in situations of unequal power, often by fathers, brothers, husbands, and lovers, all too often in drunken or drug-induced rages. In the social and political world as well, the anger of the powerful has often been turned against those who lack the power to fight back. Women have also experienced the destructive force of women's anger turned on self and other. Clearly, anger is not a simple matter. It can heal as well as harm.

Yet questions about the relation of anger and love cannot be dismissed. Feminists must ask if process philosophy's understanding of divine sympathy for human feelings can be expanded to include sympathy for our feminist anger. If the notion of the divine sympathy means that the divine power feels all the feelings of the world, then it must also be true that Goddess/God feels our anger and the hurt that caused it. In this sense our anger is in Goddess/God and will always be in Goddess/God. When the expression of anger is healing or transforming, Goddess/God is with us in our anger. When anger needs to be expressed for healing or transformation, Goddess/God will persuade us to get angry. Feminists can expand process philosophy's understanding of the divine sympathy to include sympathy for our anger. As we do so, it becomes clear that a feminist process paradigm can embrace the notion that the expression of anger can be a work of love.

However, we must always recall that in situations of unequal power relations or with force, anger can be deadly. Even when not deadly, repeated venting of angry feelings can be unproductive, closing off rather than opening dialogue. This suggests that the critical question about anger is not whether to repress it (which can be harmful to the self) or to express it (which can be harmful to others). The key is to learn how to acknowledge anger and the situations that cause it while transforming anger into forms of energy that harm neither the self nor others.[49] This means that we need to recognize that becoming aware of our anger is only the first step in a process of transformation. In the long run we need to find creative solutions to the problems that give rise to anger.

Feminists must also ask the next question. If our anger can be creative, what about the anger of Goddess/God? Can it be creative, too? Most of us are familiar with biblical images of the anger of God. In Exodus, God cast the horses and the horsemen of the Pharaoh into the sea so that the Hebrew slaves could escape their pursuers. The prophets said that God turned his anger against his own people, allowing the cities and towns of Israel and Judah to be destroyed because the Hebrew people sacrificed on mountaintops and ignored the cries of the widow at the gate. Christians imagine that God will unleash his righteous anger on the wicked on Judgment Day. In our own times, oppressed people have appealed to divine anger as legitimating their struggles for liberation. In each of these contexts, in different ways, divine anger is understood to be an aspect of divine concern for justice.

At least since classical times, divine anger has been a controversial subject. Classical theists denied that God could have any feelings at all, since to feel is to change, and God does not change. They were particularly concerned that biblical images of the divine anger suggested that God was moody and capricious. In contrast to classical traditions, process philosophy affirms that Goddess/God feels the feelings of the world with the greatest sympathy imaginable. At the same time, process philosophy insists that the divine power is persuasion, not coercion, power with, not power over. In the Bible and in Judaism and Christianity, divine anger is most frequently imagined to include the power to destroy the wicked. Such power is a clear example of power over, the power of a tyrant God. If what is meant by divine anger is the power to destroy the wicked, then it is clear that process philosophy cannot affirm it, because process philosophy denies that divine power is coercive. In other cases the anger of the biblical God is described as having been provoked by jealousy. Often the Hebrews who worship other divinities are compared to adulterous wives or prostitutes who sleep around. This sort of anger is possessive, egocentric, and expressed in terms of a husband's ownership of his wife. Theologically it is associated with exclusive rather than inclusive monotheism. Such anger is also not compatible with the process view of the divine sympathy.[50] This leaves open the question of whether there is another sense in which we would want to affirm that Goddess/God expresses anger in the form of persuasion or power with. This possibility is consistent with process philosophy, as long as it clear that divine anger is not intended to intimidate or harm but is a moment in a process of creative transformation.

"Then what," it might be asked, "does process philosophy have to offer the poor who have placed their hope in the power of God to vanquish their oppressors?" This hope is realistic only if we understand the power of God as power over rather than power with. As discussed in chapter four, the power to intervene in history with a mighty arm is not the kind of power Goddess/God has. Goddess/God is with the poor and the oppressed in their suffering. Goddess/God wants all of us to enjoy life and to increase the possibilities of enjoyment for others. Goddess/God is always persuading us to co-create a more joyful world for all people and all beings in the web of life. This is the liberating power of Goddess/God. Goddess/God can empower us to change the world, but Goddess/God cannot change it for us. In the long run, understanding divine power as power with may be more liberating than the hope that a dominating other will come in glory to destroy the wicked and vindicate the oppressed.

Building on biblical stories in which God acts to punish the wicked and liberate the oppressed, liberation theologians speak of God's concern for those who are suffering as a "preferential option" for the poor. Many Christian feminists speak of God's preferential option for poor women, the poorest of the poor.[51] This brings us to the next feminist question. Can process philosophy affirm God's preferential option for poor women? The answer to this question is yes and no. Goddess/God shares the suffering of poor women and profoundly desires their liberation. However, the idea that God has a preferential option for the poor is an extension of the biblical idea of chosenness. This idea is expressed in the Bible as the notion that God chose the Hebrew nation out of all the nations in the world to be his special people. In the understanding of many Christians, God then chose the Christian people to carry out his mission in the world. For many Americans, God next chose the American people to bring freedom and democracy to the world. Islam alleges that the followers of the Koran are God's chosen ones.

The notion of chosenness has been criticized in many contexts. Viewing one's own group as chosen by God creates feelings of national, ethnic, and religious superiority, fuels ignorance and intolerance, and stifles criticism and self-criticism. Yet it has not been adequately understood that the notion of God's preferential option for the poor arises from and perpetuates the idea of chosenness. Naomi Goldenberg has pointed out that in gatherings of religious feminists, Christian feminists (who themselves are middle or upper middle class) often use the notion of God's preferential option for the poor to assert the moral and intellectual superiority of their own version of Christian feminism.[52] I do not believe we need to speak of divine preference of any kind in order to assert that Goddess/God cares deeply about those who are suffering and profoundly desires their liberation. Process philosophy understands the sympathy of Goddess/God to be wide and deep enough to take in the whole world. The notion that God prefers one group to others has gotten us into enough trouble already. I cannot imagine why feminists would want to perpetuate it. In this case, a feminist process paradigm cautions feminists to be more careful when using biblical metaphors and habits of thinking.

In a feminist process ethics of liberation, ethical action would be inspired by feeling the feelings of others and based in an understanding that life in the body and the world is meant to be enjoyed by all individuals in the world. It would take account of the social construction of reality and the institutional character of oppression. It would name human beings as responsible for creating structures of justice and injustice. Yet it would also recognize that we are not alone in our suffering, our joy, and our struggles to create a better world. The divine power is always present, encouraging us to co-create a more joy-filled world. I believe this is the only firm basis for spiritual feminist ethics, as I discussed in the previous chapter.

In this chapter I have suggested that a feminist process paradigm offers well-thought-out alternative to dualistic anti-body, anti-female, and ultimately matricidal theologies rooted in classical theism. A feminist process paradigm affirms changing life while holding together both sides of the classical dualisms—embodiment and creative freedom, immanence and transcendence, the one and the many, female, male, and all other individuals. Process philosophy's panentheism and its notions of the divine sympathy and divine power as power with can help us to re-imagine the divine in the world. In these and other ways, a feminist process paradigm can help us to deepen our criticism of traditional ways of understanding the divine and the world and stop us from unconsciously reverting to traditional habits of thought. As we make connections between feminism and process thinking, we need not limit ourselves to placing process thinking in a feminist context or feminist thinking in a process context, and adding up the similarities and the differences. A feminist process paradigm is a new creative synthesis. Those who are involved in other spiritual paths, including Christianity, Judaism, Buddhism, Hinduism, Taoism, and indigenous religions will develop this synthesis somewhat differently than I do here. I hope that as the feminist process paradigm becomes more widely understood and used, it will help feminists in religion to recognize how much we hold in common.

In this chapter, I have suggested some of the ways that a feminist process paradigm might help us to re-imagine the divine in the world. Another test of a philosophy is the light it sheds on the process of living. Diane Martin describes how process philosophy can take root in a life. "Process is not only a philosophy, theology or speculative metaphysics," she writes, "it is a way of looking at things, a perspective on living reality, changing and evolving moment by moment. Amidst the multitude of possibilities, there is choice. Yesterday while feeling depleted and morose and pondering what lies ahead of me, I paused. I thought of the possibilities of perspective available to me as a palette of colors. Then I remembered that Goddess and God is love and at the heart of everything, always luring us to good things, well being, creativity, and expansion. I understood that I can choose consciously to identify with and embody that which is most creative—welcoming in the love that surrounds and is inside this world of vast intelligence and ever-creating movement. I too create, I co-create with the Goddess and the world. She creates and co-creates as She too unfolds and comes into being. As my life is an unfolding process, I unfold with the Goddess, the Goddess unfolds through me. I find this way of thinking about my life intelligent, liberating, creative, and comforting."[53]

Something like this has been true in my life as well. There was a time when, admiring the birds on my balcony, I might have thanked God or Goddess for making blue tits. Now when I think of the existence of blue tits as the result of millions of years of evolution in which a myriad of individuals have played a part, my wonder is far greater. If I think of my life so far in terms of what I have accomplished, I can feel sad and depressed. There are many things I have wanted to achieve yet have not. But if I think of my life as a process unfolding, I see it differently. It is more important that my life has been interesting and that I have been passionately interested in life, than that I have achieved or not achieved certain goals. When I reflect on the fate of the earth and the rate at which human beings are destroying it, I imagine that "the end" is near. But then I remember that the creative process of life is continuing and that as long as there is life, there can be joy. I redouble my efforts to increase the possibilities for joy in the world. "She changes everything She touches and everything She touches changes. Change is. Touch is. Everything we touch can change." Where there is life, there is hope.

Chapter Nine

Re-imagining Symbols

Reflecting on his career as a philosopher, Charles Hartshorne acknowledged that "a purely philosophical religion has serious limitations. There have to be symbols."[1] Philosophical reflection expresses and shapes understanding, but it is through symbols—including images, prayers, songs, dances, movements, meditations, and rituals—that the insights of the mind become part of the feelings of the body and can be shared in community. Hartshorne agreed with feminists that a process understanding of the divine in the world cannot be expressed using traditional symbols of God as a dominating male other. When a feminist process paradigm makes it clear that the body and the changing world, relationship and interconnection have been undervalued in classical theism in part because they have been identified with the female side of the classical dualisms, it becomes apparent that re-imagining the divine as female is necessary. Precisely because femaleness has been identified with the body, nature, and relationship, female images of divinity have the metaphoric power to break the hold on the human imagination of traditional images of God as a dominant male living in a heaven separate from this earth.[2] At the same time, female images for divine power should not be limited to expressing the stereotypically feminine side of divinity, for divine power is ultimately unified. There is also a need for new male images of divine power, modeled on process philosophy's understanding of power with—for example, God imaged as a father delighting in his child's growing independence and freedom.

The process of re-imagining symbols is sometimes thought to be frivolous and self-indulgent. Some have characterized rituals affirming the female body through images of Goddess, Sophia, or Shekhina as narcissistic navel-gazing, while others have expressed concern that rituals celebrating female sexuality might be excessive.[3] Sometimes this judgment is stated in a less extreme way. Re-imagining symbols is important, it might be said, but taking social, political, and ethical action against injustice is more important. Yet this is a false dichotomy. Symbols of God as a dominant male other are inspiring both acts of terrorism and preparations for wars of revenge as I write. Symbols of God as a dominant male other and notions of infallible revelation are fueling campaigns to control female sexuality around the world. In America, Africa, Asia, Europe, Australia, New Zealand, and the Pacific islands, women and girls submit to male violence every minute of every day because religions have taught them to believe that if God is male, then the male is God. While not precluding the need for other forms of social, political, and ethical action, the process of re-imagining symbols is itself a social, political, and profoundly ethical act. It is one of the ways we transform the world.

Process philosophy's understanding that every moment is a new creative synthesis of the past can help us to demystify the process of symbol creation. Traditional religious authorities have taught us that religious symbols are revealed by divine powers. Others have suggested that symbols arise from the unconscious and that every form of conscious intervention in their creation is doomed to failure.[4] Yet historical research on biblical and nonbiblical traditions reveals that religious symbol systems are always created through a process known as "syncretism," in which elements of previously existing religions are shaped into a new creative synthesis. From this perspective, the feminist process of "re-imagining," shaping new symbols using the resources of existing traditions, is not new. Ancient Hebrew, early Christian, Muslim, Hindu, Buddhist, and all other traditions were formed by creative re-imagining of existing religious symbols from new perspectives. Thus feminist re-imagining is not a departure from the way symbols have always been created, but a continuation of it. What may be different is that feminists are conscious of our part in the process of symbol creation. While we may understand that new symbols are inspired by our experiences of Goddess/God, feminists who engage in symbol creation acknowledge as well that symbols are created by human beings. Because we know the potential of religious symbols to heal and harm, we also recognize the need to ensure that the symbols that we create enhance and affirm life.

As we involve ourselves in the process of creating new images of God-She and Goddess, a feminist process paradigm can help us to evaluate the symbols that suggest themselves to our imaginations. As we plumb traditions and tap our creative powers, it is useful to have a framework that can help us to recognize life-affirming imagery and to explain (or defend) its use to others. For example, the notion of divine sympathy suggests the appropriateness of personal imagery for divine power and of prayer directed to a divine power understood as with us in every moment of our lives. The idea that the universe is the body of divinity makes it fitting to use female, male, and other than human imagery—including animal, cellular, and mineral imagery—to express the understanding that all is in Goddess/God and that Goddess/God is in the whole world. The process understanding of divine power as power with, not power over, can help us to guard against unconsciously re-introducing images of domination, including those derived from any hierarchical or violent tradition, into our feminist re-imaginings of symbols of the divine in the world. The insight that life is meant to be enjoyed calls us to create rituals that celebrate a co-created world. Let us see how this works in practice.

In the introduction, I asked whether it is appropriate to use the titles Lady and Queen to refer to divine power as female. From a process perspective, these images, like the more common Lord and King of the Bible and traditional liturgies, reflect the model of power over. A feminist process paradigm can help us to understand why it is not enough simply to change the gender of images of domination. Because traditional and modern prayers that refer to Goddess or God-She as Lady or Queen do not express the process understanding of divine power as power with, further transformation of traditional imagery is necessary.

Marcia Falk's metaphor "Source of Life," suggesting a spring or a fountain, the waters that nourish life, is an attractive image of power with. The Source of Life may also call to mind the evolutionary process in which life began and for millions of years developed in the sea. However, Marcia Falk became dissatisfied with the image of the Source of Life, because for her it implied a separation between divine power and the world. In one of her later prayers, Falk writes, "The breath of my life will bless,"[5] without naming the object of the blessing. In this prayer the world is alive with divinity. In Hebrew the phrase "breath of life" alludes to the divine spirit. Praying without separating ourselves and the world from the divine spirit is a powerful way of expressing the process understanding that the world is the body of divinity. Yet, from a process point of view, the combination of prayers to a divine power as Source of Life with others that invoke the breath of life is more complete than either would be alone.[6] In process philosophy the divine power is present in and not radically divorced from the world, but it is separate from other individuals in the sense that all individuals have a degree of creative freedom that is our own. For this reason, images naming divine power as separate from yet related to other individuals are appropriate. This separation is not absolute or radical, and it does not create the relation of dominant and dominated, because Goddess/God is internally related to every individual in the world. In feeling the feelings of every individual in a world that is the body of Goddess/God, the divine power is closer to us than we are to the cells of our own bodies. Thus, images in which there is no separation between the divine power and the world are also fitting. Insofar as none of the images that Falk chooses express the divine sympathy, the notion that Goddess/God cares about our lives and the fate of the world, from a process point of view Falk's images need to be supplemented by more personal yet non-dominating imagery.

In the controversial prayer associated with the first Christian feminist Re-Imagining Conference, the divine power was invoked as "Sophia," a female name for the divine power that means "wisdom." Though "Sophia" might not have been a familiar name to all Christians, feminists had long been searching the Wisdom literature of the Bible and tradition for female images of God.[7] However, as this imagery also comes through patriarchal traditions, it too needs to be criticized and transformed.[8] The imagery that surfaced at the Re-Imagining Conference was a product of contemporary feminist creativity reshaping the resources of the past. From a process point of view, this activity is perfectly appropriate. For process philosophy, personal imagery is fitting because it expresses the divine sympathy for the world. Female imagery is also appropriate, because process philosophy rejects the dualistic understanding of transcendence as male and immanence as female. For process philosophy, the divine power can be imagined as having a body, for the world is the body of Goddess/God. The image of Sophia's body wet with milk and honey, reminding us of pleasures and sensations, exquisitely expresses the process understanding that the divine power wants us to enjoy life in our bodies.

The image of the Asian Buddhist Goddess Kwan Yin, already known to Asian and Asian American feminists and adopted by some Caucasian feminists, is another appropriate image for the divinity of process philosophy. Traditionally Kwan Yin is invoked in the chant "Namo Guan Shih Yin Pu-sa." This can be translated as "I call upon the bodhisattva who sees and hears the sufferings of the world."[9] Reverence for Kwan Yin comes from a strand of Buddhism that some would call theistic or perhaps panentheistic. From a process point of view, the image of Kwan Yin seeing and hearing the sufferings of the world beautifully evokes the process understanding of the divine power sympathetically feeling the feelings of the world. Kwan Yin is an important reminder that traditions other than European can be resources for images of the divine as female. Yet to the extent that the image of Kwan Yin suggests that the divine power shares our suffering but not our joy, her image needs to supplemented by images such as that of Sophia giving birth or joyously making love. When we understand that life is meant to be enjoyed, we can also call on Kwan Yin share our joy and to support our efforts to lessen suffering in the world.

Some Goddess feminists have found that the Hindu image of fierce Kali with her tongue sticking out, swords in her hands, and skulls on her belt, ready to battle the forces of evil, encourages western women to express anger about injustice and the restrictions of their lives. Contemporary Indian feminists have also used the image of Kali to promote feminist interests.[10] These women may have misunderstood or rejected some understandings of Kali found within traditional Hinduism.[11] However, the issue I wish to consider here is the creative use western women have made of the image of Kali in constructing feminist spiritual alternatives. If, as I have suggested, Goddess/God can express anger as part of a transformative process, then the image of angry Kali is a fully appropriate symbol of divinity. Her image may serve a useful purpose, as feminists insist that it does, of encouraging women to get in touch with repressed anger. Yet divine anger should never be imagined as expressing itself through violence, because violence is the method of power over. In contemporary India, militant Hindu women have invoked the Hindu Goddesses in support of aggressive Hindu fundamentalism and nationalism.[12] This calls attention to the need to creatively re-imagine images of the Goddess inherited from Indo-European warrior traditions. In India as well as in Europe, the most ancient Goddess traditions were re-imagined when warrior tribes speaking Indo-European languages conquered traditional peoples.[13] As feminists continue the process of re-imagining Kali, her tongue—like that of her sister the ancient Greek Gorgon—would still be sticking out but the swords and trophies of battle would be removed. If, as some argue, the skulls on Kali's belt are not war trophies, but emblems of the prehistoric death Goddess, then they might be retained in a new (or older) context, not associated with the warrior's sword.

Kali also appeals to western feminists because her dark skin provides an inclusive alternative to the "whiteness" of God and of some of the Goddesses of European traditions. Insofar as traditional images of God are of an Old White Man, it is as important to reclaim Goddess/God as black or dark as it is to reclaim Goddess/God as female. Yet we should not identify dark-skinned Goddesses exclusively with qualities such as death and anger (as opposed to life and compassion), even when these qualities are understood positively. According to Marija Gimbutas, white was the color of death in Old European traditions, while black was the color of the fertile and transforming earth.[14] Re-imagining Kali in conjunction with other dark-skinned Goddesses such as the African water goddess Oshun, who represents fertility or creativity, sexuality, and love, provides a more complex image of the divine as female and dark-skinned.[15]

Feminists have also found the symbolism expressed in tarot cards, Goddess amulets, runes, and other forms of divination inspiring. Yet process philosophy asserts that there is no way of knowing the future until it actually is created by a multiplicity of wills. Then what is the meaning of divination? Perhaps it is this: whether we are asking about a love affair, a pregnancy, a job, the fate of a project, or the fate of the earth, no method of divination can tell us with certainty what will happen. On the other hand, the future will be a synthesis of things that already exist. Methods of divination can be understood as other than rational ways of getting a perspective on what already exists and as ways of imagining what we and others can create out of what already exists. There are people who are skilled in seeing possibilities in our lives that we may not see ourselves, or who are able to sense the desires of our hearts that we hide even from ourselves. Yet we must always take their advice with a grain of salt, for there is no one who can tell us what choices we will or should make, nor is there is anyone who knows what choices others will make until they actually do make them.

Having discussed a number of important ways other feminists are re-imagining the divine in the world, I would now like to share some of the symbols I find meaningful in my life. I have discovered and shaped images, prayers, and rituals in many years of co-creative practice of feminist Goddess spirituality. The symbols I will discuss express the feminist process understanding of the divine in the world articulated in this book. These symbols have roots in a variety of traditions, and thus will be of interest to feminists practicing diverse spiritual paths. I know that we also need new images of the divine as male, but for me the need to discover and co-create healing images of the divine as female has been more pressing. I encourage others to continue to search for and co-create healing images of the divine as male.

An ancient and enduring expression of women's spiritual creativity is the creation of home altars.[16] Choosing items to place on a home altar and arranging them in a pleasing way is a literal expression of the process understanding that creating religious symbolism involves making a new synthesis of elements of the past. As they are not under the supervision of religious authorities, home altars are a way for women to express our own understandings of the sacred. Often items from different religious traditions are placed together in ways that blur boundaries. On the altar near me as I write, I set a terra cotta image of a seated snake Goddess from Neolithic Crete; a brass owl with blue protecting eyes and two brass owl babies; a porous gray volcanic stone with a red center shaped like a vagina from the island of Lesbos where I live; and a small plasticine statue of Saint Francis of Assisi. The Neolithic Goddess from Crete is the "mascot" of the Goddess pilgrimages to Crete for women that provide a spiritual grounding in my life. She sits proudly on ample buttocks and has thick, folded, snakelike arms and legs and a beaked face.[17] She symbolizes the creative powers of divinity as source and sustainer of the world that is the divine body. Owls are an image of the Bird Goddess, one of the most ancient symbols of the Goddess.[18] Animal imagery can be used for the Goddess when the world is understood to be the divine body. Birds have long been admired for their powers of flight. Owls hunt at night, reminding us that death is part of life and that gaining wisdom does not have to be imagined as only "en-light-enment," but can also be understood as learning to trust the darkness. The volcanic stone from my island suggests the explosive and transforming powers that lie within the earth, which is a part of the body of the Goddess. Red, the color of the blood of birth and menstruation, has long been associated with the Goddess; women's blood also has transformative power. An image of Saint Francis once stood in my grandmother's garden. Though his stigmata come from another age and an understanding of the divine in the world that I do not accept, the love of Saint Francis for the birds that sit on his arms shows that he intuitively understood himself to be a child of the earth. His image reminds me of the need to find new images of the divine in the world as male. My altar is on a wide window ledge. It is framed by terra cotta roofs, green plants and trees, the gray blue Aegean Sea, and the silhouette of the coast of Turkey. Just outside the window, blue and great tits, sparrows, chaffinches, and collared doves come to take seeds from the bird-feeders. I have images of the Goddess and of the earth as the image of the body of the Goddess in other places in my home. It is important for me to live in proximity to my neighbors (though I don't get along with all of them) and in contact with the beauty of the natural world (I am lucky enough to be able to do so). The small altar on my windowsill extends out to my balcony, across to my neighbor's homes and gardens, and beyond to the Aegean Sea and the coast of Turkey shaped by the same volcanic explosions as the island of Lesbos. In my imagination, my altar includes the whole universe that is the body of Goddess/God.

Following a suggestion made by Justine and Michael Toms of New Dimensions Radio,[19] I try to remember to pray on the first of every month, knowing that my prayer will connect with those of others who are praying on the same day. In my prayers, addressed to Goddess or Goddess/God, I begin with myself, giving thanks for life and love and asking for what I need. Though asking the divine power for help might be labeled "selfish" in some traditions, I have learned from Greek women who light candles in local churches that thanking the divine power and asking for help is one of the most fundamental of all religious gestures. It expresses a deep knowing that divine power is with us in the joys and sorrows, the desires and struggles of our daily lives. A tyrant God might not have time to participate in my life or yours, but the sympathy of Goddess/God as understood in process philosophy is wide enough to include us all. After praying for myself, I enlarge the circle of my attention to include those closest to me, my cat and the dogs that sometimes share my home, neighbors, the plant and animal life in nearby gardens, and my closest friends and family. Then I name the island of Lesbos, its people, animals, plants, rocks, and rivers, asking that the environment be preserved for all who live here. Gradually I expand the circle, naming others, asking that the hungry be fed, that those who are suffering find joy, that wars cease, and that human beings learn to live in harmony with all beings in the web of life on planet earth. When I am near a shrine church I often enter it and light candles, renewing my prayers for myself, for my friends, and for the world. Following Greek custom, I approach the icon of the Panagia (Mary), whose name means "She Who Is All Holy," look into her deep brown eyes, and kiss her image. Living in Greece I have learned that the Goddess never died: I can experience her presence in the images of the Panagia. Praying in Greek churches, I express my gratitude to Greek friends whose trust that divine help is "always there" has entered into my understanding of the divine sympathy.

It is also important to me to connect to the traditions of my past.[20] I have rewritten the traditional Roman Catholic prayer known as the Hail Mary and the traditional Protestant song of praise known as the Doxology. As I am quite certain that many of the prayers we identify as Christian or Jewish are re-imaginings of pagan traditions, I find it natural to continue the process of re-imagining. Rewriting traditional songs and prayers is not a solely intellectual exercise, though the intellect does play a role. I do not sit down one day at my computer and type out alternatives to the traditional prayers and songs. Sometimes new words come to me while taking a bath or in the course of a ritual. At other times, I live in the process of re-imagining for weeks or months or years before the words feel right to me. I re-imagined the Hail Mary prayer after it entered my mind when I was standing in front of an icon of the Panagia in Crete. I sensed that it would be good to have a simple prayer that I could memorize and call to mind whenever it felt the need for it. The words of the traditional Hail Mary prayer no longer expressed my understanding of divine power. Thus I re-imagined it as Hail Goddess.[21] Others might prefer to say Hail Sophia or Hail Shekhina.

Hail Goddess full of grace!
Blessed are you,
and blessed are all the fruits of your womb.
All Holy Mother of All!
Be with us now and in the hour of our need.
Blessed be. (or: Amen)

This prayer takes me back to darkened churches where I sat with my father's mother, watching in fascination as she whispered Hail Marys and Our Fathers while touching the mauve-colored crystal rosary beads she held in her hands. As I have re-imagined it, this prayer expresses the idea that the divine power is "in" the whole world and "with us" in our daily lives. In the traditional prayer Mary is blessed for the special fruit of her womb, Jesus. In my prayer all the fruits of the womb of the Goddess, all individuals in the world, are equally blessed. Instead of asking Mary to "pray for us sinners," I ask Goddess to "be with us." According to process understanding, the divine power is always "with us," yet the purpose of prayer is to remind us of our relationship with divinity. I omitted the word "sinners" because it implies that the divine power's primary relation to our lives is judgment—reward in heaven or punishment in hell. I changed the last words of the prayer from "in the hour of our death" to "in the hour of our need" in order to underscore the fact that death is not, as classical traditions imagined, the enemy of life.

While washing the dishes or swimming in the sea and in rituals, I sometimes sing a re-imagined version of the Doxology, which I translate as the Song of Praise.[22] Here is my new version:

Praise Her from whom all blessings flow.
Praise Her all creatures here below.
Praise Her above in wings of flight.
Praise Her in darkness and in light.

The traditional hymn expressed a hierarchy of being with the "heavenly host "above" the "creatures here below" and the "Father, Son, and Holy Ghost" ruling from on high. In my version, the divine power is imagined as female. "Above" and "below" are aspects of the world that is the divine body. We are among the creatures singing praise here below, while birds, ever a delight to human beings and to Goddess/God, praise her in the flapping of their wings and in songs called out while flying. Praising her in "darkness" and in "light" affirms the ubiquity of the divine presence. Celebrating darkness as well as light is also part of a process of transforming racist habits of thinking in which darkness is associated with ignorance, sin, and evil. The melody of the Doxology when sung with gusto evokes a welling up and spilling over of gratitude from within the body. Singing it with new words, I praise the Goddess, while recalling times when I felt the presence of God in church as a child.

In the women's programs of the Ariadne Institute called the Goddess Pilgrimage to Crete and the Sacred Journey in Greece,[23] we begin each day with a series of songs and prayers that I, following Marcia Falk, call the Morning Blessing. When I remember, I say them on awakening each day. The first part of our Morning Blessing is Marcia Falk's Morning Blessing prayer, sung in English to a simple melody.[24]

The breath of my life
will bless,

the cells of my being
sing

in gratitude,
reawakening.

This song expresses gratitude to the process of life for the gift of life. Falk's understanding that the "breath of my life" will bless and the "cells of my being" sing expresses the modern scientific and process understanding that human life is not centered in the mind but in a body-mind continuum. As noted earlier, there is a resonance between the "breath" of my life and the "breath" of God that blew over the waters in Genesis. This suggests the process understanding that every day is a new creation.

The second part of the Morning Blessing is taken from Susan Griffin's prophetic ecofeminist prose poem *Woman and Nature.*[25]

This earth is my sister;
I love her daily grace,
her silent daring,
and how loved I am
how we admire this strength in each other,
all that we have lost,
all that we have suffered,
all that we know:
we are stunned by this beauty,

and I do not forget;
what she is to me,
what I am to her.

One woman says each line and the group repeats it. When imagined as a prayer to the earth understood as a symbol of the world that is the divine body, these words are filled with resonance. When I hear them under a pine tree in a group of women, surrounded by the sounds of crickets and birds and even a donkey's braying, my mind is moved back and forth as I sense that "she" who is "my sister" is the woman who sits next to me, the bird that flies above me, the donkey in the neighboring field, the scent of the cells of the pine tree, the process of life itself, and the Goddess who is in each of us and shares our joys and our sorrows with unsurpassable sympathy. In this prayer-poem, the love and understanding shared by woman and woman, by woman and nature, and (as I hear it) woman and Goddess are palpable, sensual, and erotic. In the context in which I have set them, Griffin's words powerfully express the idea that the divine power wants us to experience life deeply and to enjoy it fully—in our bodies and in the earth body.

The third part of the Morning Blessing consists of words I composed, set to a simple melody by Susan Hill.[26]

As this day dawns
in beauty,
we pledge ourselves
to repair the web.

This song moves us out into the world, naming the beauty of each day dawning as the inspiration of our desire to increase the possibilities of joy and beauty in the world through individual and communal ethical action. Following a suggestion of Marcia Falk, I wrote the prayer in the "we" form in order to remind myself that even when we pray alone, we are in community. The reference to beauty may evoke the Navajo understanding that the path of living is a Beauty Way; it expresses the process understanding that ethical action is called forth by appreciation of the beauty of life.

In the winters when I am not swimming in the sea, I practice the yoga Moon Salutation after my morning bath. It was created by a group of women who were advanced yoga teachers at the Kripalu Center for Yoga and Health in Lenox, Massachusetts. They imagined the Moon Salutation as celebrating the power of the female body and taking account of its needs, much as the Sun Salutation was created with the male body in mind.[27] The Moon Salutation Meditation, written by Laura Cornell, which I play on a tape, reminds my body of the series of stands, stretches, and squats in this incredibly powerful yoga series:

I stand tall, heart open to the world,
body full and present in all of its beauty.
(standing with arms in prayer position)
I open my arms wide to bring all of life into my being.
(opening arms and tracing the circle of the moon)
My arms form a temple above me, sheltering and protecting me.
I know that I am on holy ground.
(arms completing the circle extended with palms touching above the head)
Yielding now, softening, my body takes the shape of the crescent moon.
I see visions of women, young and old, helping and loving each other.
(bending to the side with arms still above the head and palms touching)
Rising up and bending to the other side,
I know that my softness is my strength.
I am tested, but not broken.
(bending to the other side)
Up again, I feel the sweet stillness, always present within me.
(arms above head, palms still touching)
I step wide now into a squat.
Mother Earth's ferocious powers rise up through my strong legs, hips and back.
As woman, I give birth to all that is, caring for and protecting life.
(*arms bent in "priestess" pose, legs bent and open in "birth" pose)*

Straightening arms and legs, I am a star. I am the universe.
Planets and galaxies whirl within me. I radiate in all directions.
(legs straight and spread widely apart, arms straight out to the sides)
Supple and yielding, I stretch to the side.
I open my arms and look up, opening to love and compassion.
I reach, yearning and striving, and yet rest, accepting fully.
(triangle pose)
Turning to pyramid pose, I become quiet.
Head to knee, I sense the inner workings of my own being.
(typical runners' stretch)
Lunging, I stretch long and feel the glorious length of my body.
As I look up, the moon shines on my path.
(lunge pose)
Turning now, I touch the earth,
hands on the blessed Mother, strong and steady.
Gratefully and tenderly, I bow my head.
(turning and bending to touch the earth[28])
Coming into a squat, I am connected with all animal and plant life.
My yoni open and close to the earth,
I know my body's ability to give birth, to love, to work, to pray.
I resolve to hold all of these activities as sacred.
(full squat)

The Moon Salutation continues with the poses repeated in reverse order to form a complete circle and cycle of the moon with the whole body.[29] The combination of words and yoga movement creates connections between the body and the mind, enabling the meaning of the words to come into the body. The full meaning of the Moon Salutation can be appreciated only in the doing. It celebrates the female body and the earth body, affirming that the female body is sacred, an image of the body of Goddess. It names the connection between women and the moon, positively affirming cycles of change, in contrast to classical theological traditions. In the Moon Salutation, women's changing bodies and the process of giving birth become images of the divine creativity of the Goddess. The Moon Salutation celebrates strength as supple and yielding, yet ferocious in the protection of life. These are images of strength as power with, not power over. In the Moon Salutation, the female body is not perceived negatively as it is in traditions associating femininity with the "weaker" light of the moon. Still, it might be asked: Does the Moon Salutation limit women to the body or the traditional roles associated with it? I do not find this to be so. In the Moon Salutation the female body is an image of all the creative powers in the universe. It can expand to include planets and galaxies. The female body is celebrated not only for its capacity to give birth, but also for its ability to love, to work, and to pray. Though I have not given birth, I do not feel excluded or limited by the Moon Salutation Meditation. My body is the source of all my creative powers. Nor do I imagine that as a woman I can only practice the Moon Salutation but not the Sun Salutation. For me practicing the Moon Salutation with the Moon Salutation Meditation is one moment in my spiritual practice as process. I find it, like the Sophia prayer of the Re-Imagining Conference, a powerful way of affirming the female body as an image of the creative powers of the divine body.

In a world where war is all too often seen as the solution to international problems, I stood alone in the rain at the edge of the sea on February 2, the holiday known as Candlemas, a time of new beginnings. I prayed for peace, re-imagining a familiar blessing. The original benediction reflects the understanding of divine power as power over. In it, the priest or minister asks the Lord to bless us and keep us, the light of His face to shine upon us, and Him to give us peace. It envisions divine power "out there," like the light of the sun illuminating a dark planet. Identifying divine power with light but not darkness, it continues racist habits of thinking. Attributing the power to give peace to God, it denies the roles played by human choice in the creation of both war and peace. In the re-imagined benediction, the community asks the divine power as wisdom (Sophia) to dwell within us. It understands divine power as power with it, inspiring us to co-create peace on earth.

May Goddess bless us and keep us.
May her wisdom dwell within us.
May we create peace.[30]

Through these diverse examples of re-imagining symbols of the divine in the world, I have suggested some of the ways that the insights of a feminist process paradigm can be transformed into the body knowing through prayer, ritual, song, and movement. Philosophical insight must be expressed in symbols that involve the body and community in order for it to take root in our lives. Process philosophy helps us to understand re-imagining symbols as a process of making new creative syntheses from the resources of the past. The process of re-imagining symbols is ongoing. Though we will discover and create new symbols that can work for ourselves and in communities over periods of time, we should not imagine that we will ever get our symbols "right" once and for all, for this would be to stop the process of creative re-imagining. As we creatively re-imagine symbols, it is important to remember that symbols are not an end in themselves. When they work, religious symbols express our understanding of the meaning of life, open our hearts to its wonder, and kindle our passion to co-create a world in which joy abounds for ourselves, for Goddess/God, and for all individuals in the world.

Notes

Preface

1. See Charles Hartshorne, "Do Birds Enjoy Singing? (An Ornitho-Philosophical Discourse)," *The Zero Fallacy and Other Essays in Neoclassical Philosophy,* ed. Mohammad Valady (Chicago and La Salle, Illinois: Open Court, 1997), 43–50. Also see Charles Hartshorne, *Born to Sing: An Interpretation and World Survey of Bird Song* (Bloomington: Indiana University Press, 1992 [1973]).
2. "Preface," *The Zero Fallacy,* x.
3. See Emily Culpepper, "Philosophia: Feminist Methodology for Constructing a Female Train of Thought," *Journal of Feminist Studies in Religion* 3/2 (Fall 1987): 7–16.
4. These words are taken from a Methodist hymn written by Charles Wesley, familiar in Hartshorne's childhood and in mine.

Introduction

1. "She Changes Everything She Touches," words and music by Starhawk in Starhawk, *Dreaming the Dark: Magic, Sex, and Politics,* fifteenth anniversary edition (Boston: Beacon Press, 1997), 226.
2. While the phrase "women's spirituality" is sometimes used to refer to those who have left (or never been part of) biblical traditions, in this book I will use the phrases "spiritual feminism" and "spiritual feminists" inclusively, to refer to all women who are engaged in creative re-imagining of religion from a feminist perspective.
3. A word used to refer religions indigenous to or rooted in a specific land, such as Native American, Aboriginal Australian, and many other tribal religions.
4. "Theo-logy," from *theos* (God) and *logos* (meaning) is generally understood as reflection on the meaning of God. Because God is generally

imaged as male in western traditions, Naomi Goldenberg coined the word "thea-logy" to refer to reflection on the meaning of *thea*, the female divine power or Goddess. See her *The Changing of the Gods: Feminism and the End of Traditional Religion* (Boston: Beacon Press, 1979.

5. See Carol P. Christ and Judith Plaskow, eds., *Womanspirit Rising: A Feminist Reader on Religion* (San Francisco: Harper & Row, 1979, 1989), and Judith Plaskow and Carol P. Christ, eds., *Weaving the Visions: New Patterns in Feminist Spirituality* (San Francisco: Harper & Row, 1989).
6. *Rebirth of the Goddess: Finding Meaning in Feminist Spirituality* (New York: Routledge, 1998 [1997]).
7. As evidenced in the journal *Feminist Theology*, there is more conversation among Christian, Jewish, and Goddess feminists in Great Britain than there is in the United States.
8. I suspect that one of the most significant reasons that spiritual feminists do not make common cause is institutional pressure (with financial consequences) on Christian and Jewish feminists not to associate themselves publicly with ideas, symbols, and individuals considered "pagan" or "heretical." See Nancy J. Berneking and Pamela Carter Joern, eds., *Re-Membering and Re-Imagining* (Cleveland: The Pilgrim Press, 1995), and Rita M. Gross, "Feminist Theology: Religiously Diverse Neighborhood or Christian Ghetto?" in *Journal of Feminist Studies in Religion* 16/2 (Fall 2000), 73–78 and responses 77–131, including my own, 79–84. Also see Naomi Goldenberg, "Witches and Words" and my companion piece "The Goddess and Her Cultured Despisers, " *Feminist Theology*, forthcoming.
9. See Carol P. Christ, "The Serpentine Path," *SageWoman* 56 (Winter 2001–2002), 53–56.
10. Susanne Scholz finds blurring of religious boundaries particularly evident among younger women; see "Theologizing from the Interstice: A German-Diasporic Perspective," *Journal of Feminist Studies in Religion* 18/1 (Spring 2002), 87–91.
11. I use "process philosophy" as a shorthand for the thinking of Charles Hartshorne and Alfred North Whitehead and their followers. Other process philosophers have developed the thinking of Whitehead and Hartshorne differently than I do here, yet there is general agreement on the basic outlines of process philosophy. This book is more a conversation with Hartshorne than with Whitehead, and as I have noted, like Hartshorne, I came to many of the ideas known as process philosophy on my own. I am not addressing the developments of process thought in this book, as my purpose is to make the central insights of process thinking more widely accessible and to make a connection between process and feminism.
12. I am writing this book in a dialogical style in which questions and comments—such as this one—from hypothetical readers are included

in the text. When set off by phrases like "some might say" or "you might be thinking," I will put quotation marks around these hypothetical questions or comments, but they will not be attributed to any particular source.

13. Anyone who becomes interested in process thinking will probably be directed to Whitehead; Marjorie Suchocki describes her "abortive" first reading of Whitehead in "Openness and Mutuality in Process Thought and Feminist Action," in Sheila Greeve Daveney, ed., *Feminism and Process Thought: The Harvard Divinity School/Claremont Center for Process Studies Symposium Papers* (New York and Toronto: The Edwin W. Mellen Press, 1981), 62–82.
14. This idea should be attributed to Hartshorne, who finds it expressed in Plato's *Republic,* but not to Whitehead; see *The Zero Fallacy,* 91.
15. For a definition of inclusive monotheism, see Marcia Falk, "Notes on Composing New Blessings" in *Weaving the Visions,* 128–138; also see *Rebirth of the Goddess,* 109–112; and Laurel C. Schneider, *Re-Imagining the Divine: Confronting the Backlash against Feminist Theology* (Cleveland: Pilgrim Press, 1998), chapter seven.
16. See Mary Daly, *Beyond God the Father: Towards a Philosophy of Women's Liberation* (Boston: Beacon Press, 1973); Penelope Washbourn, *Becoming Woman: The Quest for Wholeness in Female Experience* (New York : Harper & Row, 1977); Sheila Greeve Davaney, ed., *Feminism and Process Thought;* Catherine Keller, *From a Broken Web: Sexism, Separation, and Self* (Boston: Beacon Press, 1986) and *Apocalypse Now and Then: A Feminist Guide to the End of the World* (Boston: Beacon Press, 1996); Rita Nakashima Brock, *Journeys by Heart: A Christology of Erotic Power* (New York: Crossroad, 1992); Rita Nakashima Brock and Rebecca Ann Parker, *Proverbs of Ashes: Violence, Redemptive Suffering, and the Search for What Saves Us* (Boston: Beacon Press, 2001); Marjorie Suchocki, *In God's Presence: Theological Reflections on Prayer* (St. Louis: Chalice Press, 1993) and *The Fall to Violence: Original Sin and Relational Theology* (New York: Continuum, 1995); Sallie McFague, *The Body of God: An Ecological Theology* (Minneapolis: Fortress Press, 1993), *Super, Natural Christians: How We Should Love Nature* (Minneapolis: Fortress Press, 1997), and *Life Abundant: Rethinking Theology and Economy for a Planet in Peril* (Minneapolis: Fortress Press, 2001); Nancy R. Howell, *A Feminist Cosmology: Ecology, Solidarity, Metaphysics* (Amherst, NY: Humanity Books, 2000); also see my *Rebirth of the Goddess.* This list is not exhaustive. These works have varying degrees of explicit reflection on the relationships between feminist and process thought.
17. Mary Daly subtitled *Beyond God the Father "A Feminist Philosophy of Religion"* because she no longer considered it rooted in Christian confession; in that book she presents Whitehead as a resource for feminist thinking. A process-influenced reworking of the medieval Thomistic understanding of

"being" as a "Verb," as "be-ing," and as "becoming" can be discerned there and in her subsequent works. Feminist philosophy of religion has a foothold in Britain. See, for example, Pamela Sue Anderson, *A Feminist Philosophy of Religion* (Oxford: Blackwell Publishers, 1998); Grace M. Jantzen, *Becoming Divine: Towards a Feminist Philosophy of Religion* (Bloomington and Indianapolis: Indiana University Press, 1999); Beverley Clack, *Sex and Death* (Oxford: Oxford University Press, 2002). Jantzen, 255–259, 263, deals positively if briefly with the process view, yet is critical of its "realist" assumption that a divine power "exists." In the end she proposes "pantheism" for her view of the divine, not the process "panentheism."

18. See the work of Elisabeth Schüssler Fiorenza, for example *Sharing Her Word: Feminist Biblical Interpretation in Context* (Boston: Beacon Press, 1998). Schüssler Fiorenza sees no need for ontology (philosophical reflection on the nature of human or divine being or life).
19. Judith Plaskow responds to this point of view in "The Right Question is Theological," in Susannah Heschel, ed., *On Being a Jewish Feminist* (New York: Schocken, 1983), 223–233.
20. Starhawk, *The Spiral Dance: A Rebirth of the Ancient Religion of the Great Goddess* (San Francisco: Harper & Row, 1979), 25.
21. Rita Gross disagrees but had to justify her Buddhist feminist theology against this view. See *Buddhism after Patriarchy: A Feminist History, Analysis, and Reconstruction of Buddhism* (Albany: State University of New York Press, 1993).
22. See *Rebirth,* xiii-xvii.
23. His name was Joseph Botand Blazak and the course was at Citrus Junior College in Southern California in the summer of 1964.
24. See Marcia Falk, *The Book of Blessings: New Prayers for Daily Life, the Sabbath, and the New Moon Festival* (San Francisco: HarperSanFrancisco, 1996 [Boston: Beacon Press, 1999]); also see discussion xvi-xvii.
25. Marcia Falk, "The Morning Blessing," *The Book of Blessings,* xviii-xix, 10–11. [c] 1996 by Marcia Falk. Used by permission of the author.
26. Hartshorne criticizes Tillich in *Creative Synthesis and Philosophic Method* (La Salle: Open Court Publishing Company, 1970), 148–158.
27. See "A Controverted Conference," *The Christian Century,* Vol. 111 (February 16, 1994), 160–161. See *Re-Membering and Re-Imagining,* 19–20, for the words of the complete prayer which was written by Hilda A. Kuester.
28. See, for example, "The Charge of the Goddess," and "The Charge of the Star Goddess" in Starhawk, *The Spiral Dance* (San Francisco: Harper & Row, 1989 [1979]), 91–92; also see "The Hymn to Ishtar" as quoted in Anne Baring and Jules Cashford, *The Myth of the Goddess: Evolution of an Image* (London: Viking Arkana, 1991), 193.
29. Western women's attraction to Kali is widespread. When I taught in the Women's Spirituality Program at California Institute of Integral Studies in 1994, a poster of Kali decorated the common area; also see China Gal-

land, *The Bond between Women: A Journey to Fierce Compassion* (New York: Penguin Riverhead, 1998).

30. See Sandy Boucher: *Kwan Yin: Buddhist Goddess of Compassion* (Boston: Beacon Press, 1999).
31. This view is too widespread to require or benefit from being individually documented; while I was writing this book, a friend responded to a minor setback in my life with the words, "As my guru says, 'everything happens for good.'"
32. Z Budapest included feminist interpretations of traditional Tarot cards in her *Feminist Book of Lights and Shadows,* edited by Helen Beardwoman (Los Angeles: Susan B. Anthony Coven #1, 1975), subsequently published as *The Holy Book of Women's Mysteries* (Oakland, CA: Wingbow, 1989); the *Motherpeace Tarot Cards* created by Vicki Noble and Karen Vogel have been wildly successful, as have the *Amulets of the Goddess* by Nancy Blair.
33. See for example, Jacques Derrida, *The Margins of Philosophy,* trans. Alan Bass (Chicago: University of Chicago Press, 1982 [1972]); Michel Foucault, *The Archaeology of Knowledge* (London: Tavistock Publications, 1972 [1969]).
34. Among feminists in religion, Sheila Davaney, Kathleen Sands, and Sharon Welch are most frequently associated with this view. See Laurel C. Schneider, *Re-Imagining the Divine,* 76–79 for a discussion of their views, and 147–152 for a response to them.
35. Technically theology implies reflection on the nature of divine power, but some (for example Rita Gross) have expanded the meaning of the word to include reflection on the nature of reality in traditions understood to be nontheistic, such as Buddhism.
36. This would be true for example of the work of Christian feminist and womanist thinkers such as Rosemary Radford Ruether, *Sexism and God-Talk: Toward a Feminist Theology* (Boston: Beacon Press, 1983); Elisabeth Schüssler Fiorenza, *In Memory of Her: A Feminist Theological Reconstruction of Christian Origins* (New York: Crossroad, 1983); Delores Williams, *Sisters in the Wilderness: The Challenge of Womanist God-Talk* (Maryknoll NY: Orbis, 1993); as well as the Buddhist theology of Rita Gross and my Goddess thealogy, *Rebirth of the Goddess.*
37. See Alfred North Whitehead, *Religion in the Making* (New York: The World Publishing Company, 1971[1926]), 120.
38. Whitehead recognized this; see *Religion in the Making,* 123–126.
39. As I have done in *Rebirth of the Goddess,* see chapter 2, 31–49; see Laurel C. Schneider, *Re-Imagining the Divine,* 83–108, and 147–153 for a general discussion of the role of experiences of divine presence in feminist theologies and thealogies.
40. A conversation with Mary Grey alerted me to the necessity of addressing the impediments to a conversation between feminism and process before beginning the substantive arguments of this book.

41. Alfred North Whitehead, *Process and Reality,* corrected edition, edited by David Ray Griffin and Donald W. Sherburne (New York: The Free Press, 1978 [1929]); Whitehead's reflections on God can be found in a short and extremely dense concluding chapter, 342–351. Also see *A Key to Whitehead's Process and Reality,* ed. Donald W. Sherburne (Chicago and London: University of Chicago Press, 1966).
42. Alfred North Whitehead, *Modes of Thought* (New York; The Free Press, 1966 [1938]), 174.
43. Charles Hartshorne, *Omnipotence and Other Theological Mistakes* (Albany: State University of New York Press, 1984); *Wisdom as Moderation: A Philosophy of the Middle Way* (Albany: State University of New York Press, 1987); *The Zero Fallacy* has already been documented in an earlier note.
44. *Religion in the Making* (New York and Cleveland: Meridian Books, 1971[1926]).
45. See, for example, the end of *Religion in the Making,* 153.
46. Hartshorne reported that when he mentioned Hitler in conversation with Whitehead, "it was as though a shadow had been cast over his face as he replied, 'I'm afraid mankind is going into a decline.'" See *The Zero Fallacy,* 27.
47. In addition to works mentioned in previous notes, see John B. Cobb, Jr. and David Ray Griffin, *Process Theology: An Introductory Exposition* (Philadelphia: Westminster Press, 1976); John B. Cobb, Jr., *A Christian Natural Theology: Based on the Thought of Alfred North Whitehead* (Philadelphia: Westminster Press, 1965), *God and the World* (Philadelphia: Westminster Press, 1969); David Ray Griffin, *Reenchantment without Supernaturalism: A Process Philosophy of Religion* (Ithaca and London: Cornell University Press, 2001).
48. See John B. Cobb, Jr., "Feminism and Process Thought: A Two-Way Relationship," in *Feminism and Process Thought,* 34.
49. See *The Zero Fallacy,* x.
50. See *Omnipotence,* 99–103; *Wisdom as Moderation,* 59–60, 125.
51. *Omnipotence,* 58.
52. Charles Hartshorne, *The Darkness and the Light: A Philosopher Reflects upon His Fortunate Career and Those Who Made It Possible* (Albany: State University of New York 1990), 399. Also see *Omnipotence,* 60. But Hartshorne was not an essentialist; see his last published work, *The Zero Fallacy,* 22–23.
53. *Omnipotence,* 60.
54. See for example, *Omnipotence,* 44, 79, 93; *Wisdom,* 92; *The Zero Fallacy,* 48.
55. I first argued this point in "Why Women Need the Goddess," published in *Heresies 5* (1978) *(The Great Goddess Issue),* 8–11; reprinted in *Womanspirit Rising,* 273–287, and scores of times since then. Also see *Rebirth,* 22–25.

56. Hartshorne was deeply influenced by his mother's love and his father's liberal understanding of Christianity and by many other writers and philosophers including the American Philosopher C.S. Pierce, whose papers he edited, before he came into contact with Whitehead. See *The Darkness and the Light.*
57. I entered Stanford University in 1963 on a California State Scholarship; at that time the California community college, state college, and state university systems made the dream of a college education more widely attainable than ever before and possibly since. My graduate work was funded by Woodrow Wilson and Danforth fellowships.
58. In his adult life, Hartshorne and his family practiced as Unitarian Universalists. Initially, Unitarians denied the doctrine of the Trinity, while Universalists denied the doctrine of hell. Contemporary Unitarian Universalism supports a variety of spiritual paths including Christian and Pagan. Unitarian Universalists have played an important role in American religious life. I base my comment that Unitarianism lacks a lively religious symbolism of its own on discussions with many Unitarians, participation in Unitarian worship, and on Hartshorne's remark in his autobiography that "a purely philosophical religion has serious limitations. There have to be symbols." This is followed by rumination on the fact that he no longer believes in the God of the Bible, yet he wonders what he might have believed had he not been brought up believing in biblical religion. I sense a certain wistfulness in this passage that I interpret as a desire for new symbols. See *The Darkness and the Light,* 210.
59. My doctoral thesis, "Elie Wiesel's Stories: Still the Dialogue" (Yale University), was completed in 1974 and is available from University Microfilms.
60. My leaving the church was as much about the statements that the Jews rejected Christ in the Bible and the Roman Catholic Easter mass in which I had participated for a number of years as it was about the sexism of religious language. See *Laughter of Aphrodite: Reflections on a Journey to the Goddess* (San Francisco: HarperSanFrancisco, 1987), 7–10.
61. *Diving Deep and Surfacing: Women Writers on Spiritual Quest* (Boston: Beacon Press, 1980, 1985, 1995). The projected chapter on Denise Levertov was not written.
62. See previous note.
63. Through the Ariadne Institute for the Study of Myth and Ritual, www.goddessariadne.org.
64. *Odyssey with the Goddess: A Spiritual Quest in Crete* (New York: Continuum, 1995).
65. *The Zero Fallacy,* 43–50.
66. Each one played a particular role. Cristina is widely read and interested in nonwestern thinking and in the Goddess. She read each chapter in various stages and her questions and insights significantly affected the flow of my thought. Judith read the manuscript through her commitment to

rigorous thinking and ethically engaged Judaism. John was effusive with praise and kept me from making mistakes in my interpretations of process thinking. With each of the three of them interested in my work, I felt I had everything I needed. Indo-European linguist and Goddess feminist Miriam Robbins Dexter also read all of the chapters and provided support. I have been lucky to have had wonderful editors. Gayatri Patnaik encouraged this project before I began it and urged me to address it to the widest possible audience; Kristi Long provided insightful comments on the next-to-last draft, and Michael Flamini helped to bring the book to completion. Nancy Filson also read drafts of many of the chapters, and Marcia Hilton listened and asked questions as I told her how the first chapters of the book were developing in our afternoons by the sea. I am also grateful for readings by Rita Gross and by an anonymous Christian feminist process theologian to whom the manuscript was sent by the publisher.

67. See Charles Hartshorne, *Creative Synthesis and Philosophic Method.*
68. See note 1 in this chapter.

CHAPTER ONE

1. In Christian Orthodox iconography the archangels still wear the armor of medieval knights.
2. Mary Daly, "After the Death of God the Father," in *Womanspirit Rising,* 53–62.
3. Hartshorne calls his philosophy of religion "neoclassical theism" in order to distinguish it from "classical theism."
4. See Hartshorne, "Pantheism and Panentheism" in *The Encyclopedia of Religion,* Vol. 11, ed. Mircea Eliade (New York: Macmillan, 1987), 168.
5. These professors were Hans Frei and Jaroslav Pelikan.
6. Elie Wiesel, *The Gates of the Forest,* trans. Frances Frenaye. (New York: Avon Books, 1967).
7. Martin Buber, *I and Thou,* trans. Walter Kaufmann (New York: Scribners', 1978 [1958]). While retaining the title used by the earlier translator, Kaufmann argued that in English "thou" sounds stilted and that "you" is more intimate and therefore the appropriate translation of "du," the informal "you." Also see *I and Thou,* Ronald Gregor Smith, trans. with a postscript by the author (New York: Collier Books, 1987 [1958]).
8. Charles Hartshorne, *The Divine Relativity: A Social Conception of God* (New Haven and London: Yale University Press, 1948).
9. Charles Hartshorne, *Omnipotence,* 1–49. Whereas Hartshorne discussed God's unsympathetic goodness after omnipotence and omniscience, I discuss it second because it so clearly follows from God's unchangeable-

ness. I consider the mistakes in yet another order in subsequent chapters of this book and therefore I will refer only to the first theological mistake by number.

10. Plato, *The Symposium,* trans. Walter Hamilton (Baltimore: Penguin, 1951), 93–94. Plato attributes this speech to Socrates, who says he learned the ideas from a woman of Mantinea named Diotima.
11. *The Divine Relativity,* 25.
12. *Omnipotence, 10–12.*
13. *The Zero Fallacy,* 75.
14. Women, it was thought, might sit at the feet of the men and listen, or as one less generous rabbi put it, they might function as "footstools" for the men.
15. Rosemary Radford Ruether has repeatedly written that the possibility of life after death is at best unknown and should not be a matter of concern for Christians. See *Sexism and God-Talk, 235–258;* and *Gaia and God: An Ecofeminist Theology of Earth Healing* (San Francisco: HarperSanFrancisco, 1992), 251–253.
16. See Elizabeth Stuart, "A Good Feminist Woman Doing Bad Theology?" *Feminist Theology* 26 (January 2001), 70–82, esp. 79–82.
17. See Starhawk, M. Macha Nightmare, and the Reclaiming Collective, *The Pagan Book of Living and Dying: Practical Rituals, Prayers, Blessings, and Meditations on Crossing Over* (San Francisco: HarperSanFrancisco, 1997), 78–97.
18. See *Laughter of Aphrodite* (San Francisco: HarperSanFrancisco, 1987), 213–227, for the first view; and *Rebirth of the Goddess* (New York: Routledge, 1998), 132–134, for the second.

CHAPTER TWO

1. From here on I will be using "Goddess/God" to refer to the divine power as conceived in process thought (including my interpretations of it), but will retain the word "God" to refer to the deity intended by classical theism. See the introduction, 17–18, for a discussion of this term.
2. See Whitehead, *Modes of Thought* (New York: The Free Press, 1968 [1938]), 127–147.
3. See Marija Gimbutas, *The Language of the Goddess* (San Francisco: Harper & Row, 1989) for the view that in the religion of Old Europe in the Neolithic era the processes of birth, death, and regeneration were revered. This understanding influenced the Eleusinian Mysteries that were practiced in the time of Plato.
4. Scholars allege that the Genesis creation stories do not in fact imagine creation out of nothing. In the Genesis 1 creation story, the spirit (or

breath or wind) of God blows over apparently pre-existing waters. In the Genesis 2 story, God molds apparently pre-existing earth or clay.

5. In describing traditional views, I follow tradition in using "man" (with added quotation marks) as generic. To do otherwise would in many cases be distorting. Traditional theologians did not always mean to include women when they said "man." I put quotation marks around "woman" when referring to traditional views.
6. All of the so-called higher religious traditions (Judaism, Christianity, Hinduism, Buddhism, Islam, and Confucianism) are to greater and lesser degrees influenced by ascetic thinking. This means that they all have elements of animosity or ambivalence toward women, toward the body, and most probably toward nature as well. Unless recognized and criticized, these attitudes may be consciously and unconsciously transposed into present day religious thinking. Many westerners, feminists among them, have been introduced to Hinduism or Buddhism or Chinese philosophy by modern teachers or writers who downplay the ascetic strain of their traditions in order to attract a modern western audience. When attitudes toward women and the body that have been fueled by asceticism are later discovered in sacred texts or teachings, they may seem confusing.
7. In focusing on evolution, I am following Hartshorne in *Omnipotence,* ch. 3. Whitehead was concerned with reconciling philosophy with quantum physics; see *Science and the Modern World* (New York: The Free Press, 1925), ch. 8.
8. See Pierre Teilhard de Chardin, *The Phenomenon of Man,* trans. Bernard Wall (New York: Harper & Row, 1958), and Brian Swimme and Thomas Berry, *The Universe Story: From the Primordial Flaring Forth to the Ecozoic Era—A Celebration of the Unfolding of the Cosmos* (San Francisco: HarperSanFrancisco, 1992).
9. See *Omnipotence,* 135–136.
10. Charles Hartshorne, *Creative Synthesis,* xv, 57–68; also *Omnipotence,* 22.
11. *Omnipotence,* 67.
12. Thus sociobiologists argue that males will inevitably seek multiple partners in order to spread their genes around; see, for example, Richard Dawkins, *The Selfish Gene* (Oxford and New York: Oxford University Press, 1989). Even apparently socially motivated behavior will be explained away as really being about individual genetic interests, as in the case of the "altruistic" birds discussed in the next chapter.
13. See Alfred North Whitehead, *Process and Reality,* 18 and following.
14. See *Omnipotence,* 60–63; *Wisdom,* 21–22; *The Zero Fallacy,* 132–150.
15. *Wisdom,* 22;
16. *The Zero Fallacy,* 138.
17. *Omnipotence,* 62.
18. *Creative Synthesis,* 6.
19. See *Omnipotence,* 120.
20. *Omnipotence,* 79.

21. When he coined the terms panpsychism and panpsychicalism, Hartshorne was arguing that matter has soul in contrast to the view of modern science that matter is lifeless. However, it would be a mistake to read Hartshorne as saying that soul can exist without body. This is one of the cases where a feminist perspective can push process philosophy into a clearer affirmation of and expression of embodiment and the body-mind continuum.
22. For example Rachel Carson, *The Edge of the Sea,* with a new introduction by Sue Hubbell (Boston and New York: Houghton Mifflin Company, 1998 [1955]) and *The Sea Around Us* (Oxford and New York: Oxford University Press, 1989 [1950]); also Jennifer Ackerman, *Chance in the House of Fate: A Natural History of Heredity* (Boston and New York: Houghton Mifflin Company, 2001).
23. A copy of this letter, handwritten, not dated but sent in the late spring of 1998, is now on file at the Center for Process Studies in Claremont, California.
24. *Wisdom,* 24.
25. Martin Buber, *I and Thou,* 58.
26. And this certainly seems to be true in my garden.
27. See, for example, Jeffrey Moussaieff Masson and Susan McCarthy, *When Elephants Weep: The Emotional Lives of Animals* (New York: Delacorte Press, 1995); Stephen Hart, *The Language of Animals* (New York: Henry Holt, 1996); Alexander F. Skutch, *The Minds of Birds* (College Station: Texas A & M University Press, 1996); Frans de Waal, *Good Natured: The Origins of Right and Wrong in Humans and Other Animals* (Cambridge, MA: Harvard University Press, 1966); and Hartshorne's own *Born to Sing.*
28. Joanne Harris, *Chocolat* (London: Black Swan, 2000).
29. See *Omnipotence,* 52–56.
30. See J. E. Lovelock, *Gaia: A New Look at Life on Earth* (Oxford: Oxford University Press, 1982).
31. *Omnipotence,* 79.
32. See Charles Hartshorne, "Pantheism and Panentheism," in the *Encyclopedia of Religion,* Vol. 11, 165–171. Hartshorne attributes the word "panentheism" to the German philosopher K. F. Krause (1781–1832).
33. See *Omnipotence,* 44–49, and elsewhere.
34. The terms "primordial" and "consequent" were used by Whitehead, while Hartshorne said "abstract" or "absolute" and "concrete" or "contingent." Hartshorne's view differs slightly from Whitehead's; I will be presenting Hartshorne's view of dual transcendence in this book.

CHAPTER THREE

1. See Charles Hartshorne, *The Divine Relativity.*
2. See Plato, *The Symposium,* 93–94.

3. I now realize that I had trouble understanding the logic behind "social contract" theory because I had never assumed that human beings could exist apart from a web of personal and social relationships. As usually happens in the educational system, I thought this theory was something that I just didn't "get."
4. Martin Buber, *I and Thou,* 78.
5. Ntozake Shange, *for colored girls who have considered suicide when the rainbow is enuf* (New York: Macmillian, 1976).
6. This line of thinking reached its apogee with Leibniz, whose theory of the monads will be discussed below.
7. Remember the paper called "A New Proof for the Existence of Other Minds," mentioned in the chapter one.
8. And indeed almost exactly a year later, the first black stork to arrive was feeding at the far edge of the reservoir.
9. Hartshorne attributed this view to Charles Pierce and Henri Bergson; see Charles Hartshorne and Creighton Peden, *Whitehead's View of Reality* (New York: Pilgrim Press, 1981), 8.
10. *Whitehead's View of Reality,* 8.
11. For a discussion of Leibniz by Hartshorne, see *The Zero Fallacy,* 134–136; for the importance of Leibniz to Whitehead, see Alfred North Whitehead, *Process and Reality,* 19.
12. See *Omnipotence,* 104–109.
13. *Omnipotence,* 105.
14. See Paula Gunn Allen, *The Sacred Hoop: Recovering the Feminine in American Indian Traditions* (Boston: Beacon Press, 1989), 141.
15. See David Abrams, *The Spell of the Sensuous: Perception and Language in a More-than-Human World* (New York: Vintage Books, 1997), 163–172.
16. The word "suffering" once had a wider meaning in the English language than it does today.
17. See Chris Mead, "Surrogate Parents," *Bird Watching* (August 2001), 22–24
18. *The Divine Relativity,* x.
19. See my *Laughter of Aphrodite,* 21.
20. See my *Odyssey with the Goddess,* 60.
21. See *Odyssey,* 25–26.
22. This phrase from a Methodist hymn of the same name by Charles Wesley was often quoted by Hartshorne as expressive of the process view of Goddess/God. See *The Zero Fallacy,* 167.
23. The allusion is to words attributed to the apostle Paul who said in Romans 8:38–39 that "I am sure that neither death, nor life, nor angels, nor principalities, nor things present, nor things to come, nor powers, nor height, nor depth, nor anything else in all creation, will be able to separate us from the love of God in Christ Jesus our Lord."
24. This term and the understanding of it was proposed by Whitehead and adopted by Hartshorne; see Hartshorne, *The Divine Relativity,* 142. Whitehead also speaks of the divine persuasion as "luring."

CHAPTER FOUR

1. Starhawk in *Dreaming the Dark* offers the alternative of "power-from-within" to "power-over"; Carter Heyward speaks of "power-in-relationship" and "mutuality" in *The Redemption of God: A Theology of Mutual Relation* (Washington, DC: University Press of America, 1982); Riane Eisler contrasts "domination" and "partnership" in *The Chalice and the Blade: Our History, Our Future* (San Francisco: Harper & Row, 1987); I use "power with" and "power over" to describe the process view; Whitehead and Harshorne more commonly speak of "persuasion" and "coercion."
2. See Martin Buber, *I and Thou.*
3. See Hartshorne, *Omnipotence,* 58.
4. See for example, H. H. Rowley, *From Moses to Qumran* (New York: Association Press, 1963), 141–183; Samuel Terrien, *Job: Poet of Existence* (Indianapolis: Bobbs-Merrill, 1957); and Theodore Friedman, "Job, The Book of," *Encyclopedia Judaica,* Vol. 10 (Jerusalem: Keter Publishing House, Ltd., 1972), 122–123.
5. *Omnipotence,* 73, quoting the English clergyman Charles Kingsley.
6. On the other hand, it might make sense to question overreliance on automobiles.
7. I recognize that Hinduism can be understood in both monistic and non-monistic ways. It is generally understood by philosophers to be monistic. Yet Hindu practices of devotion to divinities or divinity suggest that some Hindus may have implicitly polytheistic, theistic, or panentheistic understandings of ultimate reality.
8. Hartshorne defines worship as loving God with one's whole heart, mind, and soul.
9. The notion of God as the "ground of being" is taken from Paul Tillich, but I add "becoming" to make it clear that being is always in process. This phrase is not found in Hartshorne. In addition, my understanding of the ground of being differs from Tillich's in that I agree with Hartshorne that dual transcendence enables us to say both that Goddess/God is the ground of all being and becoming and that Goddess/God is the supreme or surrelative being who enters into relationship with individuals in the world. See Hartshorne on Tillich, *Omnipotence,* 32, 47.
10. *Omnipotence,* 71.
11. *Omnipotence,* 67.
12. The relationship between the abstract and unchanging and concrete and changing aspects of Goddess/God should not be imagined as a linear sequence, that is, it should not be understood as suggesting that first Goddess/God existed, then Goddess/God created the values and principles of being and becoming, and then Goddess/God created the first

possible world. As stated earlier, process philosophy does not imagine a time when Goddess/God was not related to some possible world.

13. See Job 42:6.
14. This is the epithet of Kwan Yin, the Asian Goddess of compassion; see Sandy Boucher, *Discovering Kwan Yin.*
15. *Omnipotence,* 75.
16. See Jonathan Schell, *The Fate of the Earth* (New York: Avon Books, 1982).
17. Speaking of the nuclear threat, Hartshorne wrote: "Thus the perilous 'experiment of nature,' a species as free from instinctive guidance as ours, is approaching a critical stage. Was the experiment too dangerous? . . . if I play at criticizing God it is at this point." See *Omnipotence,* 126.
18. *Omnipotence,* 58.

CHAPTER FIVE

1. Alice Walker, *The Color Purple* (New York: Pocket Books, 1983), 178.
2. See Mary Grey, "Gender, Justice and Poverty in Rural Rajasthan—Moving beyond the Silence," *Feminist Theology* 25 (September 2000), 33–45.
3. See Charles Hartshorne, *Born to Sing.*
4. Lillian Lawler, *The Dance in Ancient Greece* (Middletown, CT: Wesleyan University Press, 1964).
5. Rachel Carson, *The Edge of the Sea,* 250.
6. "Introduction" by Sue Hubbell, *The Edge of the Sea,* xx. Also see Linda Lear, *Rachel Carson: Witness for Nature* (New York: Henry Holt and Company, 1998 [1997]).
7. Doreen Valiente crafted "The Charge of the Goddess" in which these words occur, using a model found in Charles Leland's *Aradia;* see *The Rebirth of Witchcraft* (Custer, WA: Phoenix Publishing Company, 1989), 61–62; also see Starhawk, *The Spiral Dance,* 91–92.
8. Conversation reported to me by Rita Nakashima Brock.
9. See Charles Hartshorne, *The Darkness and the Light,* 207–208.
10. Ruth Mantin, conversation.
11. *The Zero Fallacy,* 46–47.
12. In the United States, the domination known as Universalists who believed in God but not in hell joined with the Unitarians, who believed in God but not in the Trinity, to form the Unitarian Universalist Association. Unitarian Univesalists embrace all spiritual paths.
13. See Elizabeth Kubler-Ross, *On Death and Dying* (London and New York: Tavistock Publications, 1970 [1969]).
14. Christian process theologian David Ray Griffin comes to a different conclusion about the possibility of life after death; see *Reenchantment,* 230–246.

15. *Omnipotence,* 35.
16. This view is held by many Christians and also by many philosophers, including the existentialists.
17. To paraphrase the apostle Paul (Romans 8: 39–39).

Chapter Six

1. Andrew Harvey tells the story of his devotion to Guru Ma in *Hidden Journey: A Spiritual Awakening* (New York: Henry Holt, 1991).
2. Gerald Gardner, the founder of modern Witchcraft, created the fiction that Wiccan tradition was passed on in secret from pre-Christian times; see Ronald Hutton, *The Triumph of the Moon: A History of Modern Pagan Witchcraft* (Oxford: Oxford University Press, 1999), chapter eleven.
3. See Starhawk, *The Spiral Dance,* 91–92.
4. See Doreen Valiente, *The Rebirth of Witchcraft,* 57–62.
5. See *Odyssey with the Goddess.*
6. This story is told in *Odyssey with the Goddess.*
7. This is also the assumption of popular Twelve Step Programs for alcohol and other addictions.
8. See Valerie Saiving, "The Human Situation: A Feminine View," in *Womanspirit Rising,* 25–42.
9. This comment was made by Judith Plaskow some years ago following a session at the American Academy of Religion.
10. *Process and Reality,* xiv.
11. For a consideration of these questions by a feminist philosopher who is in dialogue with enlightenment and post-enlightenment philosophical traditions, see Pamela Sue Anderson, *A Feminist Philosophy of Religion.* Also see my discussion of embodied thinking in *Rebirth of the Goddess,* chapter two.
12. This is the phrase used by Madhu to describe our relationship. I support her through the Nepalese Youth Opportunity Foundation; see chapter seven.
13. See my *Laughter of Aphrodite,* chapters one and two.
14. The phrase is Nelle Morton's.
15. Among the groups I founded or helped to found are the Yale Women's Alliance, a graduate student consciousness-raising and direct action group; the Yale Faculty, Staff, and Student Alliance, the group that brought the Department of Health, Education, and Welfare to investigate Yale for sex discrimination (and found it guilty); a consciousness-raising group in New Haven; the Association of Graduate Student Women in the Religious Studies Department at Yale; the Women's Caucus—Religious Studies and the Women and Religion Section of the

American Academy of Religion; the New York Feminist Scholars in Religion; the women's ritual group that eventually took the name Rising Moon.

16. See *Diving Deep and Surfacing* (Boston: Beacon Press, 1995 [1980]); the projected chapter on Denise Levertov was never written.
17. See *Diving Deep and Surfacing,* 62.
18. See Starhawk, *The Spiral Dance;* and Z Budapest, *The Feminist Book of Lights and Shadows* and *The Holy Book of Women's Mysteries,* Vol. 1.
19. Traditions influenced by Witchcraft or Wiccan traditions as created by Gerald Gardner.
20. This phrase is often used by Wiccans in invoking the four directions. Ronald Hutton has traced it to Masonic tradition, see Ronald Hutton, *The Triumph of the Moon.*
21. See *The Triumph of the Moon.*
22. See *Laugher of Aphrodite,* chapter eleven.
23. See *Odyssey with the Goddess.*

CHAPTER SEVEN

1. Though the title of this chapter is the same as that of *Reason for Hope: A Spiritual Journey* by Jane Goodall with Phillip Berman (New York: Warner Books, 1999), it also forms part of the subtitle of a section of *Rebirth of the Goddess,* 155.
2. Rachel Carson, *The Silent Spring,* introduction by Al Gore (Boston/New York: Houghton Mifflin, 1994 [1962]), 5.
3. Susan Griffin, *Woman and Nature: The Roaring Inside Her* (New York: Harper & Row, 1978), 226.
4. Epigraph to *The Silent Spring,* v.
5. *Reason for Hope,* 232.
6. Katy Payne, "Among the Apes," *New York Times Book Review,* October 24, 1999, 21.
7. The version of the story I heard concluded with the assertion that even monkeys on other islands knew how to crack open coconuts. I do not include this line because it seems to me to imply a kind of magical effect in the changing of conscious that I find implausible.
8. I am using the term "liberal" as it was used in early modern theories to refer to the emphasis on individual freedom, not as it is used in contemporary U.S. politics to refer to members of the Democratic party as social welfare theorists.
9. See *Rebirth of the Goddess,* 167.
10. John and Anne Bowers founded Friends of Green Lesbos, www.greenlesbos.com.

11. These friends are Judith Plaskow, Martha Acklesburg, and Alexander Plaskow Goldenberg.
12. Their addresses are: Nepalese Youth Opportunity Foundation, 3030 Bridgeway, Suite 211, Sausalito, California, 94965, www.nyof.org ; Global Fund for Women, 1375 Sutter Street, Suite 400, San Francisco, California 94109, www.globalfundforwomen.org; Greenpeace International, Keizersgracht 176, 1016 DW Amsterdam, Netherlands, www.greenpeace.org; Greenpeace USA, 702 H Street NW, Suite 300, Washington DC 20001, USA, www.greenpeaceusa.org; Greenpeace Greece, Zodoxou Pigis 52y, Athens 10681, Greece www.greenpeace.gr; World Wildlife Fund International, Avenue du Mont Blanc, 1196 Gland, Switzerland, www.wwf.org; World Wildlife Fund, US, 1250 24th St. NW, Washington, DC 20037, www.worldwildlife.org; World Wildlife Fund, Greece, Filellinon 15, Athens 10558, Greece, www.wwf.gr.
13. Information on the programs of the Nepalese Youth Opportunity Foundation can be found on its website www.nyof.org. It was founded by Olga D. Murray and Allan Aistrope.

CHAPTER EIGHT

1. This description of She Who Changes is created in contrast to the picture of God as an Old White Man in chapter one. Echoes from the Bible, a Christian Sunday school hymn, the invocation to Kwan Yin, and the Reclaiming chant are intended.
2. This is a commonplace in feminist theological analysis; see Mary Daly, *Beyond God the Father;* Rosemary Radford Ruether, *Sexism and God-Talk;* Judith Plaskow, *Standing Again at Sinai;* Naomi Goldenberg, *Returning Words to Flesh: Feminism, Psychoanalysis, and the Resurrection of the Body* (Boston: Beacon Press, 1990); and my *Laughter of Aphrodite,* among many others.
3. Rita Gross in *Buddhism after Patriarchy* makes a similar argument. Few other feminists have advanced it, perhaps because most feminists are not interested in promoting or validating the ascetic world-view.
4. In classical Athens, the Eleusinian Mysteries dedicated to the Goddesses Demeter and Persephone were widely practiced. The Greeks recognized that these mysteries had an ancient origin; scholars have traced them back to early agricultural societies.
5. See Paul Roche, trans., *The Orestes Plays of Aeschylus* (New York: New American Library, 1962), *The Eumenides,* 190.
6. *Science and the Modern World* (New York: The Free Press, 1967 (1925), 48.

7. An allusion to Hartshorne's *Insights and Oversights of Great Thinkers: An Evaluation of Western Philosophy* (Albany: State University of New York Press, 1983).
8. See Hartshorne, *The Darkness and the Light,* 62–64.
9. See Judith Plaskow, "The Coming of Lilith," in *Womanspirit Rising: A Feminist Reader on Religion* (San Francisco: Harper & Row, 1979 (1989), 198–209.
10. See *Laugher of Aphrodite,* 126.
11. See *Weaving the Visions,* 208–213. Carter Heyward, Rita Nakashima Brock, Judith Plaskow, and I are among those who have used Lorde's ideas in feminist work in religion.
12. See Elinor Gadon, *The Once and Future Goddess* (San Francisco: HarperSanFrancisco, 1989); also see Meinrad Craighead, *The Mother's Songs: Images of God the Mother* (New York: Paulist Press, 1986), whose work has appealed to Christian, Jewish, and Goddess feminists.
13. See the Introduction.
14. Angela Farmer, *The Feminine Unfolding* (videotape) (Hohokus, New Jersey: Transit Media, 1999); also see Laura Cornell, *The Moon Salutation: Expression of the Feminine in Body, Psyche, Spirit* (Oakland, CA: Yogeshwari Publications, 2000).
15. See Rita Gross, *Buddhism after Patriarchy;* and Miranda Shaw, *Passionate Enlightenment; Women in Tantric Buddhism* (Princeton: Princeton University Press, 1994).
16. See for example, Christine Downing, *The Goddess* (New York: Crossroad, 1984), and Jean Shinoda Bolen, *Goddesses in Everywoman* (San Francisco: Harper & Row, 1984).
17. The use of this term is widespread; see, for example, Carter Heyward, *The Redemption of God.*
18. The use of this term is also widespread, see, for example, eds. Irene Diamond and Gloria Feman Orenstein, *Reweaving the World: The Emergence of Eco-Feminism* (San Francisco: Sierra Club Books, 1990).
19. See Sheila Greeve Davaney, "The Limits of the Appeal to Women's Experience, in *Shaping New Vision: Gender and Values in American Culture,* eds. Clarissa W. Atkinson, Constance H. Buchanan, and Margaret Miles (Ann Arbor, MI: University of Michigan Research Press, 1987), 31–49.
20. See Starhawk, "Power, Authority, and Mystery: Ecofeminism and Earth-based Spirituality," in *Reweaving the World,* 73.
21. Catherine Keller's *From a Broken Web* discusses the relational self from a process standpoint.
22. Reprinted in *Womanspirit Rising,* 25–42.
23. Valerie Saiving, "Androgynous Life: A Feminist Appropriation of Process Thought," in *Feminism and Process Thought,* 28. The essay was presented as the Harvard University Dudelian Lecture in 1978.
24. Grace M. Jantzen, *Becoming Divine: Towards a Feminist Philosophy of Religion* (Bloomington and Indianapolis: Indiana University Press, 1999).

25. See my *Laughter of Aphrodite,* 213–227.
26. See Rosemary Radford Ruether, *Gaia and God,* 251–253.
27. As discussed in chapter two.
28. The idea that the world is the body of God is affirmed by Hartshorne but not Whitehead, as I have said.
29. Marcia Falk, "Notes on Composing New Blessings," in *Weaving the Visions,* 128.
30. See Sallie McFague, *The Body of God.*
31. See Lynn Gottlieb, *She Who Dwells Within: A Feminist Vision of a Renewed Judaism* (San Francisco: HarperSanFrancisco, 1995).
32. Sallie McFague uses process philosophy to make this point in *The Body of God.*
33. Hartshorne addressed this question in *Creative Synthesis and Philosophic Method,* 148–158.
34. See Mary Daly, *Beyond God the Father;* and Elizabeth Johnson, *She Who Is: The Mystery of God in Feminist Theological Discourse* (New York: Crossroad Press, 1992).
35. Sallie McFague, *Metaphorical Theology: Models of God in Religious Language* (Minneapolis: Fortress Press, 1982).
36. *Re-Imagining the Divine,* 64.
37. This is the part of the ontological argument advanced by Hartshorne in which he asserts that the divine power, if it is to be affirmed at all, must be imagined as sympathetic and loving. See *The Logic of Perfection* (La Salle, IL: Open Court, 1962).
38. Marcia Falk's prayers are discussed in the Introduction and in the last chapter.
39. See, for example, Elisabeth Schüssler Fiorenza, *Sharing Her Word* (Boston: Beacon Press, 1998), 51.
40. See Judith Plaskow, *Just Sex* (New York: Palgrave/St. Martins Press, forthcoming).
41. Ronald Hutton in *The Triumph of the Moon* has documented the fallacy of alleging that there is an unbroken line in which pagan traditions have been handed down from ancient times.
42. See Rita Gross, *Buddhism after Patriarchy.*
43. *The Divine Relativity,* 25.
44. "Transcendence and Immanence," *Encyclopedia of Religion,* Vol. 15, 21.
45. See Hartshorne, "Transcendence and Immanence," 19–20.
46. Influenced by process philosophy, Rita Nakashima Brock and Rebecca Parker question the centrality of the notion of self-sacrifice as a model for understanding the crucifixion of Jesus in *Proverbs of Ashes;* they are working on a new book on the history of theories of self-sacrifice and alternatives to them, tentatively titled *Night Fire;* also see Rita Nakashima Brock's earlier *Journeys by Heart.*
47. Thanks to Judith Plaskow for prodding me to reflect on the relation of the relation anger and love here.

48. See Beverly Harrison, "Anger as a Work of Love, " in *Weaving the Visions,* 214–225.
49. Thanks to Rita Gross for this suggestion.
50. Thanks to Miriam Robbins Dexter for reminding me to mention this.
51. This view is widespread; see, for example, *Feminist Theology from the Third World: A Reader,* ed. Ursula King (Maryknoll, NY: Orbis Books), 1994.
52. See Naomi Goldenberg, "Witches and Words," *Feminist Theology,* forthcoming.
53. Diane Martin, my first student in process philosophy, personal communication, December 2002.

CHAPTER NINE

1. See *The Darkness and the Light,* 210.
2. I have discussed this point in greater detail in *Rebirth,* 22–25 and 98–101.
3. This view has been expressed by Rosemary Radford Ruether and others as a criticism of the Goddess movement. See Naomi Goldenberg, "Witches and Words," forthcoming in the British journal *Feminist Theology* 35 (January 2002), and my companion piece, "Reflections on the Goddess and Her Cultured Despisers," in the same issue.
4. This view of archetypal psychologist Carl Jung, *Psychological Reflections,* ed. Jolande Jacobi (New York: Harper & Row, 1961) and his followers was repeated in theologian Paul Tillich's highly influential book *The Dynamics of Faith* (New York: Harper & Row, 1957).
5. See the introduction.
6. In other words, I agree with Falk's decision to keep both kinds of prayers in her prayer book, but not with her later argument that imagining the divine as an individual creates separation and therefore is inappropriate, nor with her assumption that personal language for divinity is not fitting.
7. This literature is widespread; see, for example, Susan Cady, Marian Ronan, and Hal Taussig, *Sophia: The Future of Feminist Spirituality* (San Francisco: Harper & Row, 1986); Maria Pilar Aquino and Elisabeth Schüssler Fiorenza, eds., *In the Power of Wisdom: Feminist Spiritualities of Struggle* (London: SCM Press, 2000) (also available in French, Spanish, Italian, German, and Dutch).
8. For a critique of the Sophia traditions, see Pamela J. Milne, "Voicing Embodied Evil: Gynophobic Images of Women in Post-Exilic Biblical and Intertestamental Text," *Feminist Theology 30* (May 2002), 61–99.
9. See Sandy Boucher, *Discovering Kwan Yin,* 107–108.
10. See Sharanda Sugirtharajah, "Hinduism and Feminism: Some Concerns," *Journal of Feminist Studies in Religion 18/2* (Fall 2002), 103–104.

11. Rita Gross in a personal communication stated, " The severed head held by Kali is one's own, teaching us to accept our own limits and finitude. Until we can accept our mortality and finitude, we will never have peace or joy." Gross also said that I interpret the warrior aspect of Hindu Goddess imagery "too literally." While I agree with Gross that some aspects of Buddhist tradition do not take the warrior imagery literally, I am always "literal" when it comes to warrior images because I recognize their origin in warlike cultures and their potential to inspire violence and the notion that "God is on our side" when we go to war.
12. See Surigirtharajah, 104.
13. Sanskrit is an Indo-European language. In *The Language of the Goddess* Marija Gimbutas documented the existence of pre–Indo-European Goddess traditions in Old Europe. Similar processes of transformation of pre-existing Goddess symbols by Indo-European warrior cultures occurred in Europe and in India.
14. See *The Language of the Goddess.*
15. See Luisah Teish, *Jambalaya: The Natural Woman's Book of Personal Charms and Practical Rituals* (San Francisco: Harper and Row, 1985).
16. See Kay Turner, *Beautiful Necessity: The Art and Meaning of Women's Altars* (New York: Thames and Hudson, 1999).
17. There is a picture of this image in *Rebirth,* 14.
18. See Marija Gimbutas, *The Language of the Goddess.*
19. New Dimensions is an independent producer of broadcast dialogues from a variety of traditions and cultures. Its goal is to deepen connections to self, family, community, the natural world, and the planet. See www.newdimensions.org.
20. I was brought up in liberal Presbyterian churches (where predestination and divine judgment were rarely if ever mentioned), but my father's family was Roman Catholic, and in graduate school I was a "practicing" if not baptized Vatican II Catholic.
21. I am happy for others to use this prayer in rituals or liturgies, if it is attributed to Carol P. Christ. The original words of the prayer are:

    Hail Mary, full of grace; the Lord is with thee.
    Blessed art thou among women, and blessed is the fruit of thy
        womb, Jesus.
    Holy Mary, Mother of God,
    pray for us sinners now and at the hour of our death.
    Amen.

22. Again, the right to reprint this song for use in rituals or liturgies is granted, as long as it is attributed to Carol P. Christ. The original words are.

    Praise Him from whom all blessings flow.
    Praise Him all creatures here below.

Praise Him above ye heavenly host.
Praise Father, Son, and Holy Ghost.

23. Information about Ariadne Institute can be found at www.goddessariadne.org.
24. See the introduction. "Morning Blessing" by Marcia Falk, *The Book of Blessings: New Jewish Prayers for Daily Life, the Sabbath, and the New Moon Festival* (San Francisco: HarperSanFrancisco, 1996 [Boston: Beacon Press, 1999]), 10, [c] 1996 by Marcia Falk. Used by permission of the author. The music is mine. It is as follows:

25. Susan Griffin, *Woman and Nature,* 219, italics and punctuation as in the original, but I have set it as poetry.
26. Again this can be used by others, if the words are attributed to Carol P. Christ and music to Susan Hill. Music as follows:

27. I learned the Moon Salutation Meditation from Laura Cornell, used with permission of Laura Cornell. See her *The Moon Salutation: Expression of the Feminine in Body, Psyche, Spirit* (Oakland, CA: Yogeshwari Publications, 2000) for a description and discussion of the Moon Salutation. The version of the Moon Salutation Meditation given here is an earlier version of the one in her book.
28. This is a variation of a difficult pose.
29. The Moon Salutation forms a complete circle, as do the words. The rest of the words are as follows:

Re-emerging now, I touch the earth,
hands on the blessed Mother, strong and steady.
Gratefully and tenderly, I bow my head.
Lunging, I stretch long and feel the glorious length of my body.
As I look up, the moon shines on my path.

Turning to pyramid pose, I become quiet.
Head to knee, I sense the inner workings of my own being.
Supple and yielding, I stretch to the side.
I open my arms and look up, opening to love and compassion.
I reach, yearning and striving, and yet rest, accepting fully.
Straightening arms and legs, I am a star. I am the universe.
Planets and galaxies whirl within me. I radiate in all directions.
I come down into a squat.
Mother Earth's ferocious powers rise up through my strong legs, hips and back.
As woman, I give birth to all that is, caring for and protecting life.
Standing again, my arms form a temple above me, protecting and sheltering me.
I know that I am on holy ground.
Yielding now, softening, my body takes the shape of the crescent moon.
I see visions of women, young and old, helping and loving each other.
Rising up and bending to the other side, I know that my softness is my strength.
I am tested but not broken.
Up again, I feel the sweet stillness always present within me.
Through this journey I have discovered the mystery of the ever-changing moon,
and I carry her wisdom and power deep within me.
I stand tall, heart open to the world,
body full and present in all of its beauty.
I take a quiet breath, my arms floating softly down to my sides.
I stand in stillness, resting as one.

30. My version can be used if credited to Carol P. Christ. Shekhina or Sophia may be substituted for Goddess in the first line. A version of the traditional benediction is as follows:

The Lord bless you and keep you.
The Lord make His face to shine upon you and be gracious unto you.
The Lord lift up His countenance upon you and give you peace.

Index